The Art of Business Origami
Folding Your Way to Success

Lloyd Jose Fernandez

Exploring the metaphorical concept of origami to reveal
innovative strategies for business adaptation, agility, and
transformation.

Index

Navigating Uncertainty with Strategy

Chapter 5: The Third Fold - Innovative Thinking
The Origami Mindset
Key Principles of The Origami Mindset
Approaches To Fostering Innovative Thinking
Real-World Examples
Unfolding Possibilities

Chapter 6: The Fourth Fold - Adaptive Culture
Building a Culture of Agility
Core Principles of An Adaptive Culture
Approaches to Building an Adaptive Culture
Real-World Examples Of An Adaptive Culture
Unfolding Resilience and Innovation

Chapter 7: The Fifth Fold - Customer-Centricity
The Symphony of Customer-Centricity
Approaches to Customer-Centricity
Real-World Examples
Creating Customer-Centric Experiences

Chapter 8: The Sixth Fold - Resilient Operations
The Essence of Resilient Operations
Approached to Resilient Operations
Real-World Examples
The Future

Chapter 9: The Seventh Fold - Sustainable Growth
Sustaining Success Over Time
The Historical Perspective

Introduction

In the vast and ever-evolving realm of the business world, where change is the only constant, and the path to success seems as intricate as the folds of an origami masterpiece, we find ourselves at the crossroads of innovation and adaptation. Welcome to "The Art of Business Origami: Folding Your Way to Success," a journey that promises not just insight, but a profound shift in the way you perceive and navigate the complex tapestry of modern commerce. In the vast tapestry of the business world, where change is the only constant and uncertainty reigns supreme, the need for adaptation has never been more vital. The rules of the game are ever-evolving, and traditional strategies are often rendered obsolete overnight. In this ever-shifting landscape, the true leaders of industry are those who possess the audacity to rethink, reconfigure, and redefine their businesses. They are the origamists of the corporate world, deftly folding and unfolding their enterprises with precision and purpose.

Imagine this book as your personal guide through uncharted territory, a companion on a thought-provoking adventure where we'll use the metaphorical concept of origami to unveil innovative strategies for business adaptation, agility, and transformation. But don't be fooled by the conventional approach you might expect from a business book; this is a departure into the extraordinary. This is not your typical business book. It is a journey into the heart of innovation, agility, and transformation, guided by the metaphorical concept of origami—a form of artistry that transforms a flat sheet of paper into intricate, three-dimensional wonders through a series of meticulous folds.

But be warned, this book does not offer a step-by-step blueprint for success, nor does it promise a one-size-fits-all solution. Instead, it invites you to embark on a thought-provoking journey, challenging you to unfold your own unique path to success. You will be inspired, questioned, and nudged to question the status quo. The journey won't be without its challenges, but it is through these challenges that true growth and transformation occur.

In the pages that follow, we will explore how this ancient art form can serve as a powerful allegory for navigating the complexities of modern business. Each fold in the origami process represents a crucial aspect of success, from visionary leadership to strategic agility, from innovative thinking to resilient operations. Together, these folds create a multifaceted strategy for adaptation in a world where adaptation is not a luxury but a necessity.

Chapter 1, "The Folded Business World," will lay the foundation by exploring the current state of business. We'll delve into the challenges and disruptions that have become synonymous with modern commerce, setting the stage for why adaptation is not just a choice but an imperative.

Chapter 2, "The Art of Business Origami," serves as our gateway into the heart of this book. Here, we'll define the concept of Business Origami and establish its philosophy as a powerful framework for adaptation. It's a chance for you to envision how the principles of origami can be applied to your business, reshaping it into something extraordinary.

Now, let's move into Part I, where we'll start unfolding the layers of our metaphorical origami.
In Chapter 3, "The First Fold - Visionary Leadership," we explore the critical role of leadership in the origami process. Just as a skilled origamist envisions the final shape before making the first fold, a visionary leader crafts a compelling vision that inspires and guides their team.

Chapter 4, "The Second Fold - Strategic Agility," takes us into the realm of strategy. In a world where change is constant, we'll learn how to adapt and navigate the ever-shifting business landscape with strategic finesse.

With Chapter 5, "The Third Fold - Innovative Thinking," we unravel the secrets of creativity and innovation. Just as origami requires imaginative folds to create intricate forms, businesses thrive when they foster a culture of innovation.

Next, in Chapter 6, "The Fourth Fold - Adaptive Culture," we explore the importance of creating an organizational culture that embraces change as a constant. Like the pliability of paper, an adaptive culture ensures that your business can fold and unfold gracefully.

Chapter 7, "The Fifth Fold - Customer-Centricity," is all about placing the customer at the heart of your business. We'll discover how the art of folding can be applied to customer experiences, creating bonds that stand the test of time.

Chapter 8, "The Sixth Fold - Resilient Operations," teaches us the importance of operational resilience. Just as an origami structure needs to withstand external pressures, your operations should be prepared for unexpected shocks.

Finally, in Chapter 9, "The Seventh Fold - Sustainable Growth," we explore the delicate balance between growth and stability. Sustainable growth isn't just about expansion; it's about maintaining the integrity of your business as it evolves.

As we move into Part II, we'll look at "The Art of Origami in Action."
Chapter 10, "Case Studies in Business Origami," brings these concepts to life through real-world examples. We'll examine how successful organizations have applied the principles of Business Origami to achieve remarkable transformations.

However, no journey is without its challenges. Chapter 11, "The Challenges of Unfolding," addresses common pitfalls that can hinder your progress. We'll equip you with the knowledge to overcome resistance to change and navigate the hurdles that lie ahead.

In Part III, "The Origami Masterclass," we delve into the process of mastering the art of Business Origami provides guidance on becoming an origami leader. We'll discuss the importance of continuous learning and improvement, ensuring that you're always ready to adapt and evolve.

Now, as we reach the conclusion of our journey, we'll reflect on "The Endless Possibilities of Business Origami." This is where we tie everything together, emphasizing that the path to success is as unique and intricate as an origami masterpiece.

In the Appendix, you'll find additional resources and tools to aid your ongoing journey. We've also compiled references and further reading for those who want to explore these concepts in more depth.

Before we wrap up this introduction, let's take a moment to acknowledge those who've contributed to this book. The insights and wisdom shared here are the result of collaboration, mentorship, and collective knowledge. I am grateful to those who've paved the way for us to explore the art of Business Origami.

In closing, this is not just a book; it's an immersive experience. "The Art of Business Origami" invites you to embark on a journey where innovation meets adaptation, where unconventional thinking becomes your secret weapon, and where success is not just a destination but a continuous unfolding.

The Power of Metaphor in Business

In the vast landscape of business, where decisions shape destinies and strategies chart courses, it's easy to get lost in the complexities. The corporate world is often likened to a chessboard, a battlefield, or even a jungle, where only the fittest survive. Metaphors like these have become the lifeblood of how we understand and communicate about business. But what if I told you that metaphors are more than mere linguistic flourishes? What if I said that metaphors have the power to shape the very fabric of how we think, act, and succeed in business?

Welcome to the intriguing world of metaphorical thinking in business, a world where words and concepts take on new life and meaning. In this chapter, we'll embark on a journey to explore the profound impact of metaphors on our understanding of the business landscape. We'll delve into why metaphors matter, how they shape our perspectives, and why they are fundamental to the art of Business Origami.

Metaphors: More Than Words

At its core, a metaphor is a figure of speech that expresses one concept in terms of another, often unrelated, concept. It's the art of drawing connections between seemingly disparate ideas to shed light on something familiar by likening it to something else. Think about the phrase "time is money." Time and money are not the same thing, but the metaphor highlights how we value and prioritize time in the context of business.

Metaphors are everywhere in business. We talk about "climbing the corporate ladder," "thinking outside the box," and "navigating through rough waters." These metaphors aren't just colorful language; they shape our understanding of success, innovation, and adversity. They provide mental frameworks that guide our decisions and actions.

The Business Jungle

Consider the metaphor of business as a jungle. When we view business through this lens, we see it as a competitive arena, where survival depends on agility, strategy, and the ability to outmaneuver rivals. This metaphor influences how we approach competition, framing it as a battle for dominance where only the strongest thrive.

For instance, Steve Jobs, the visionary co-founder of Apple Inc., often embraced the jungle metaphor in his approach to business. He once famously stated, "Innovation distinguishes between a leader and a follower." This metaphor framed the business world as a competitive jungle where only the most innovative and adaptable would thrive. Steve Jobs' commitment to this metaphor fueled his relentless pursuit of groundbreaking technologies and products, making Apple a global leader in innovation and design. The jungle metaphor wasn't just a linguistic choice for Jobs; it was a driving force behind Apple's success.

However, the jungle metaphor isn't the only way to perceive business. If we shift to the metaphor of business as a symphony, suddenly, our perspective changes. In this view, business becomes a harmonious collaboration of diverse talents, where each instrument (employee) plays a vital role in creating a beautiful, synchronized melody (the company's success). This metaphor emphasizes teamwork, cooperation, and the value of individual contributions.

Metaphors as Mental Maps

Metaphors aren't just rhetorical devices; they're cognitive tools. They serve as mental maps that help us navigate the complexities of business. Just as a map guides us through unfamiliar terrain, metaphors help us make sense of the business world by drawing parallels to concepts we already understand.

Imagine you're entering a dense forest for the first time, armed only with a map of a city you've never visited. The city map won't be much help in the forest; you need a map that's tailored to the environment you're in. Metaphors do the same in business. They provide us with mental maps that guide our decision-making and problem-solving within the specific context of business.

The Metaphor-Reality Connection

Now, you might be wondering, "Do metaphors really have that much influence on our actions?" The answer is a resounding yes. Research in the fields of linguistics and cognitive science has shown that metaphors are not just linguistic flourishes; they actively shape our thoughts, behaviors, and even our physical experiences.

In a groundbreaking study, researchers asked participants to read a passage about crime as either a "beast" or a "virus" ravaging a city. Those who read the "beast" metaphor were more likely to advocate for aggressive law enforcement measures, while those exposed to the "virus" metaphor leaned towards strategies focused on prevention and containment. The choice of metaphor significantly influenced participants' attitudes and policy preferences.

This study highlights the power of metaphors to frame issues and guide our responses. In business, the metaphors we use can lead us to adopt specific strategies, make certain decisions, and view challenges in particular ways. If we think of business as a battleground, we might prioritize competition above all else. If we see it as a puzzle, we'll focus on problem-solving and strategy. And if we view it as a journey, we'll emphasize progress and growth.

Shaping Our Reality with Metaphors

Metaphors not only shape our perception of reality but also influence the actions we take, which in turn shape our reality. It's a continuous cycle that can either limit our thinking or expand our possibilities.

For example, imagine you're leading a team tasked with launching a new product. If you frame this challenge as a "race against time," you're likely to prioritize speed and efficiency, which can be crucial in a competitive market. However, if you view it as a "journey of discovery," you might emphasize the importance of exploration, experimentation, and learning from setbacks.

The metaphors we choose don't just describe our reality; they actively shape it. They set the tone for our mindset, guide our decision-making, and influence the culture of our organizations. This is where the art of Business Origami comes into play.

Business Origami: Unfolding New Perspectives

The concept of Business Origami invites us to explore alternative metaphors for business. Just as origami transforms a flat sheet of paper into intricate forms, Business Origami encourages us to transform our businesses using fresh perspectives and innovative metaphors.

Consider the metaphor of business as a garden. In a garden, you don't just focus on competition; you nurture growth, cultivate relationships, and create an environment where every plant (employee) can flourish. This metaphor shifts your focus from competition to collaboration and sustainability. Or think of business as a conversation. In this view, success depends on effective communication, active listening, and the exchange of ideas. The metaphor of business as a conversation places emphasis on building relationships, understanding customer needs, and adapting to feedback.

The power of Business Origami lies in its ability to help us reframe our understanding of business by choosing metaphors that align with our goals and values. It allows us to consciously shape our organizational culture, strategies, and actions by selecting metaphors that inspire the kind of business we want to create.

Crafting Your Business Metaphor

As you embark on your journey into Business Origami, consider the metaphor that resonates most with your vision and values. Do you see business as a battleground, a puzzle, a journey, a garden, or a conversation? Your chosen metaphor will serve as your guiding star, influencing your leadership style, your approach to innovation, and your organizational culture.

Throughout this book, we'll explore various metaphors and how they can be applied to different aspects of business origami, from visionary leadership to strategic agility. We'll examine how adopting specific metaphors can transform the way you think about and approach business challenges.

But here's the key takeaway from this chapter: **metaphors are not just linguistic devices**

The Unconventional Approach of Business Origami

In a world that often celebrates convention, embracing the unconventional can be a transformative choice. In the realm of business, where best practices and traditional models abound, the pursuit of innovation and adaptation demands a willingness to break free from the ordinary. It's in this spirit of unconventionality that we embark on our exploration of Business Origami—a concept that defies the status quo and invites us to unfold new possibilities in the corporate landscape.

In this chapter, we'll dive deep into the unconventional approach of Business Origami. We'll explore why it stands out in a sea of business methodologies and why its metaphorical framework offers a fresh perspective for leaders and organizations seeking to thrive in an ever-changing world. So, get ready to challenge your assumptions about how business should be done.

Beyond the Playbook

Businesses, particularly successful ones, often operate according to established playbooks and best practices. These playbooks are based on historical data, tried-and-true strategies, and industry norms. While there's undeniable value in learning from past successes and avoiding common pitfalls, there's a limit to what these conventional approaches can achieve.

Imagine a scenario where you're playing chess with a friend. If you stick to well-known opening moves and strategies, you're playing "by the book." You might win occasionally, but true mastery of the game requires thinking beyond the established plays. It involves the creativity to adapt to your opponent's moves and to create novel strategies on the fly.

Business Origami is the art of playing chess without the rulebook. It encourages leaders and organizations to break free from pre-defined strategies and embrace the flexibility of origami-like thinking. It's about creating your own playbook based on the unique challenges and opportunities you face. In essence, Business Origami encourages you to be the master of your own game.

Embracing the Uncertainty

One of the hallmarks of Business Origami is its embrace of uncertainty. Traditional business models often seek to mitigate risks and maintain stability. While these goals are essential, they can sometimes lead to rigidity and resistance to change.

Picture a river with a straight, unyielding dam constructed to control the flow of water. The dam's purpose is to ensure a constant, controlled supply of water. However, when an unexpected flood occurs, the dam's rigidity can lead to disastrous consequences.

In contrast, imagine the same river with a flexible, origami-like approach. Instead of a rigid dam, a system of adaptable gates and channels is in place. When a flood comes, the gates can be adjusted, and the water is redirected safely. This system acknowledges the inevitability of change and is designed to respond to it, not resist it.

Business Origami embraces uncertainty in a similar way. It acknowledges that the business landscape is fluid, subject to sudden shifts, and influenced by external factors beyond our control. Rather than trying to maintain an unyielding status quo, it encourages leaders to anticipate change, adapt to it, and even leverage it as an opportunity for growth.

The Metaphorical Power of Origami

The use of metaphor is a cornerstone of Business Origami, and it's here that we encounter its truly unconventional nature. While many business methodologies rely on concrete frameworks, models, and metrics, Business Origami invites us to explore the abstract and metaphorical.

Let's revisit the metaphor of origami itself. Origami is the art of folding paper to create intricate, three-dimensional shapes. It begins with a flat, often unassuming sheet of paper and transforms it into something beautiful and complex. Similarly, Business Origami starts with the fundamental elements of business—ideas, strategies, and organizations—and encourages us to fold and shape them in innovative ways.

This metaphorical approach may raise eyebrows in a world where data-driven decision-making and proven methodologies are the norm. However, the power of metaphor lies in its ability to transcend the limitations of concrete thinking. Metaphors unlock our creativity, enabling us to see familiar concepts from fresh angles and inspiring unconventional solutions.

The Limitations of Convention

Conventionality, while comforting and safe, often has its limitations. When businesses adhere too rigidly to established norms and practices, they risk becoming stagnant. They may find themselves ill-prepared to adapt to sudden disruptions, changing consumer preferences, or emerging technologies.

Consider the story of Kodak, a company that dominated the photography industry for over a century. Kodak was a pioneer in photographic technology, but its steadfast commitment to conventional film and paper-based photography became a hindrance when digital photography emerged. Despite inventing the digital camera in the 1970s, Kodak struggled to adapt to the digital revolution, and eventually, the company filed for bankruptcy in 2012.

Kodak's downfall serves as a stark reminder of the dangers of clinging to convention in a rapidly evolving world. It underscores the need for businesses to embrace an unconventional approach, one that values adaptability, innovation, and metaphorical thinking.

Beyond the Business Origami Toolkit

At this point, you might be wondering, "How can I apply Business Origami to my own endeavors?" It's a valid question because, in essence, Business Origami is more than a toolkit of strategies and tactics. It's a mindset—a way of thinking and approaching challenges that goes beyond the conventional.

In the chapters that follow, we'll explore the specific folds of Business Origami, from visionary leadership to strategic agility, from innovative thinking to resilient operations. Each fold represents a distinct aspect of this unconventional approach, offering fresh perspectives and strategies for navigating the complexities of the business world.

Business Origami doesn't provide a one-size-fits-all solution because it recognizes that each organization and leader is unique. Instead, it equips you with the tools to craft your own solutions, to fold and shape your business in ways that resonate with your vision and values.

Embracing the Art of Business Origami

As we journey deeper into the world of Business Origami, keep in mind that its unconventionality is its strength. It challenges us to look beyond the playbook, to embrace uncertainty as an opportunity, and to harness the metaphorical power of origami to innovate and adapt.

The world of business is ever-evolving, and the rules are continually rewritten. In the chapters ahead, we'll explore how Business Origami offers a flexible, adaptable, and creative approach to thrive in this dynamic landscape.

Setting the Stage for Transformation

Imagine a theater bathed in soft, dimmed lights, the stage an empty canvas awaiting the performance of a lifetime. The anticipation in the air is palpable. Now, let's shift our focus from the world of the arts to the world of business. In the corporate theater, the stage is your organization, and the performance is the transformation that can elevate your company to new heights.

In this chapter, we'll set the stage for transformation, a concept that has propelled countless companies to unprecedented success. We'll explore the practical methods and approaches that have been the driving force behind the transformation of large companies, making their stories not just inspirational but also a roadmap for your own journey.

The Call for Transformation

Every great transformation begins with a recognition—a call to action. It's a moment when leaders and organizations realize that they must evolve to survive and thrive in an ever-changing landscape. The catalyst for transformation can come from various sources: a shift in market dynamics, emerging technologies, or the recognition that the status quo is no longer sustainable.

Consider the case of IBM, a global technology giant that has undergone multiple transformations throughout its history. In the early 1990s, IBM faced a crisis as its traditional mainframe business declined. The company recognized the need for transformation and embarked on a bold journey to shift its focus to software and services. This transformation, guided by then-CEO Lou Gerstner, revitalized IBM and enabled it to remain a dominant force in the technology industry.

The key takeaway here is that transformation isn't reserved for companies in crisis. It's a proactive response to changing times and evolving customer needs. Successful transformations are driven by leaders who understand that stagnation is the enemy of progress.

The Transformational Leadership

Setting the stage for transformation begins with visionary leadership. Leaders are the directors of the corporate theater, and their vision serves as the script. They must articulate a compelling vision for the organization's future—a vision that inspires and mobilizes the entire team.

One leader who embodies transformational leadership is Satya Nadella, the CEO of Microsoft. When Nadella took the helm in 2014, Microsoft faced challenges in a rapidly evolving tech landscape. Nadella's vision was clear: to transform Microsoft from a software giant into a cloud-first, mobile-first company.

Under his leadership, Microsoft successfully shifted its focus to cloud computing, leading to remarkable growth and positioning the company as a leader in the industry. Nadella's transformational leadership was characterized by his ability to communicate a bold vision, empower his team, and foster a culture of innovation.

Creating a Culture of Change

Vision alone is not enough; it must be supported by a culture that embraces change. Just as the stage crew and actors work together seamlessly to bring a theatrical performance to life, organizations must create an environment where employees are empowered to contribute to the transformation.

A shining example of a company that fostered a culture of change is Netflix. In its early days, Netflix was primarily a DVD rental service. However, the company recognized the emerging trend of digital streaming and decided to pivot its business model. This required a massive shift in culture, from a DVD-by-mail mindset to a digital-first approach.

Netflix's leadership understood that true transformation required not just a change in strategy but a cultural shift. They encouraged risk-taking and innovation, even if it meant disrupting their existing business. This culture of change enabled Netflix to become a global streaming giant, revolutionizing the entertainment industry.

Data-Driven Decision-Making

Transformational journeys are often accompanied by uncertainty. To navigate this uncertainty, organizations must rely on data-driven decision-making. Data provides valuable insights that guide strategic choices and validate the effectiveness of transformation initiatives.

Amazon, the e-commerce behemoth, is a prime example of a company that leveraged data to drive transformation. Amazon uses data not only for optimizing its supply chain and logistics but also for understanding customer behavior and preferences. This data-driven approach led to innovations such as personalized recommendations and Amazon Prime, which transformed the way consumers shop online.

For your own transformational journey, consider how data can be harnessed to inform decisions, identify opportunities, and measure progress. Whether it's customer data, market trends, or operational metrics, data can be your compass, guiding you toward your transformation goals.

Agile Strategy Execution

Transformations are not static events; they are dynamic processes that require agile strategy execution. Just as a theater performance evolves with each rehearsal, organizations must adapt their strategies based on real-time feedback and changing circumstances.

Apple, one of the world's most iconic technology companies, is known for its ability to execute strategies with agility. Apple's transformation under Steve Jobs— from a computer manufacturer to a consumer electronics and software powerhouse—was marked by its commitment to innovation and adaptability.

Apple's product launches are meticulously orchestrated, but they are also flexible, with the company adjusting its strategies based on market feedback and emerging trends. This agile approach to strategy execution has allowed Apple to consistently stay ahead of the curve and remain a market leader.

Measuring Success

Transformation is not a nebulous concept; it should yield tangible results. Just as a theater performance is judged by the applause it receives, organizations must define clear metrics and key performance indicators (KPIs) to assess the success of their transformation efforts.

One of the most notable examples of transformational success is the turnaround of the automobile manufacturer, Ford. In the mid-2000s, Ford faced financial challenges and declining market share. Alan Mulally, the CEO at the time, implemented a transformation plan that included cost-cutting measures, a renewed focus on quality, and the introduction of innovative new vehicles.

The success of Ford's transformation was evident in its financial results. The company not only weathered the economic downturn but also became profitable again. This transformation was a testament to the power of setting clear objectives and measuring progress.

Building an Ecosystem of Partnerships

In today's interconnected world, no organization is an island. Successful transformations often involve building strategic partnerships and alliances to leverage complementary strengths and resources.

A prime example of this approach is the partnership between Starbucks and Spotify. In an effort to enhance the in-store music experience, Starbucks joined forces with Spotify to create Starbucks' own music ecosystem. This partnership allowed Starbucks to offer a personalized music experience to its customers, driving customer engagement and enhancing the ambiance of its stores.

When setting the stage for transformation, consider how partnerships and alliances can amplify your efforts. Collaborations can provide access to new markets, technologies, and expertise, accelerating your transformation journey.

The Unconventional Path to Success

As we conclude this chapter, it's essential to recognize that transformation is not a linear path, nor is it without its challenges. It demands bold leadership, a culture that embraces change, data-driven insights, agile strategy execution, and a clear measurement of success. It requires organizations to be willing to pivot, adapt, and innovate continually.

But the rewards of transformation can be extraordinary. Just as a captivating theater performance leaves a lasting impact on its audience, a successful transformation can propel your organization to new heights of success and relevance. It can ensure that your company remains not just a participant but a star in the ever-evolving business theater.

The journey may be unconventional, but it promises to be a thrilling adventure filled with challenges, triumphs, and the opportunity to redefine your organization's story. It's time to take center stage and craft a transformational narrative that will leave a lasting legacy.

Part I: Unfolding the Business Landscape

In Part 1 of our captivating journey through "The Art of Business Origami," we embark on an exhilarating expedition through the foundational chapters that serve as the cornerstone for your exploration. Just as a skilled origamist masterfully unfolds a sheet of paper, revealing its hidden dimensions, we systematically uncover the layers of the business world to unveil innovative strategies essential for adaptation, agility, and transformation.

Chapter 1: The Folded Business World

In the vast arena of business, where strategies are devised, fortunes are made, and industries rise and fall, imagine if you could peel back the layers of the corporate landscape and peer beneath the surface. What would you see? Would you find a seamless, unbroken path of progress, or would you uncover a complex and intricate tapestry, rich with challenges, opportunities, and constant change?

Picture the business world as a vast origami creation, each company a unique fold in the intricate design. Just as an origamist approaches a blank sheet of paper with vision and precision, business leaders navigate an ever-shifting landscape. In this chapter, we embark on a journey to unfold the enigmatic layers of the contemporary business world, revealing the profound intricacies that underlie its surface.

The Current State of Business

To understand where we are going, we must first comprehend where we stand. The current state of business is a reflection of the world we live in—a world characterized by unprecedented technological advancements, global interconnectedness, and a relentless pace of change.

The Digital Revolution
At the heart of this transformation is the digital revolution. The advent of the internet, mobile technology, and the proliferation of data and artificial intelligence have ushered in a new era of business. Companies that once operated within the confines of physical spaces now have the potential to reach global audiences, tapping into markets that were once unimaginable.

Consider the rise of e-commerce giants like Amazon, which has redefined retail by seamlessly connecting buyers and sellers across the world. What started as an online bookstore has evolved into a behemoth that offers everything from cloud computing services to original streaming content. Amazon's success is emblematic of the digital age, where adaptability and innovation are paramount.

The Power of Data
In this era, data is king. Every click, transaction, and interaction generates a digital footprint that can be harnessed for insights and decision-making. Companies like Google and Facebook have built empires around data, using it to refine their products and services, as well as to create targeted advertising campaigns.

The Rise of Startups
The digital landscape has also leveled the playing field, allowing startups and disruptors to challenge established industry leaders. Innovations like ride-sharing services (Uber, Lyft), short-term lodging platforms (Airbnb), and

recently with the advancements and progress of artificial intelligence technologies (OpenAI), introducing new business models that are agile and customer-centric.

Globalization

Businesses are no longer confined by geographical boundaries. The rise of globalization has led to the interconnectedness of markets, enabling companies to source talent and customers from around the world. Multinational corporations like Apple and Coca-Cola have mastered the art of global brand management, creating products and experiences that resonate across cultures.

However, this interconnectedness also means that businesses are exposed to a broader range of risks, from economic fluctuations in far-flung markets to geopolitical tensions that can disrupt supply chains.

Why Adaptation is Imperative

In the unfurling pages of business history, one theme stands out with unmistakable clarity: adaptation is not a luxury; it's a survival imperative. Businesses that fail to adapt risk obsolescence. But why is adaptation so critical in the modern business landscape?

Rapid Technological Advancements

Technology is advancing at an unprecedented pace. The devices we carry in our pockets today possess more computing power than the computers that sent astronauts to the moon. This constant innovation not only creates new opportunities but also renders existing business models obsolete. Just think of AI and Large Language Models like OpenAI GPT, Google LaMDA and Meta AI disrupted the technology industry in 2023 or how streaming services like Netflix and Amazon Prime transformed the television industry.

Shifting Consumer Expectations
Customers are more informed and demanding than ever before. In the age of instant gratification, they expect seamless, personalized experiences. Businesses that can't meet these expectations risk losing customers to competitors who can. Think about how Amazon's one-click shopping and same-day delivery have set new standards for convenience.

Global Competition
Globalization has opened doors to competition from all corners of the world. Companies that once dominated their local markets must now contend with challengers from emerging economies. This global competition compels businesses to continuously innovate and improve to maintain their competitive edge.

Evolving Regulatory Environment
The world of blockchain and cryptocurrencies is a prime example of the ever-shifting regulatory terrain. Governments and financial authorities around the globe are continuously formulating new regulations to grapple with the transformative impact of digital currencies.

For instance, the regulatory landscape in the United States regarding cryptocurrencies is in a state of constant evolution. The Securities and Exchange Commission (SEC) has been actively assessing whether certain cryptocurrencies should be categorized as securities, subjecting them to specific regulations.

Environmental and Social Responsibility
There is a growing recognition that businesses must operate with a heightened sense of social and environmental responsibility. Sustainability is no longer a buzzword but a fundamental consideration. Companies that fail to address these concerns risk reputational damage and consumer backlash.

Pandemic and Black Swan Events
The COVID-19 pandemic served as a stark reminder of the unpredictability of black swan events. Businesses that were unprepared for the sudden disruptions faced unprecedented challenges. Adaptability became a matter of survival, with companies pivoting their operations and embracing remote work to stay afloat. In the face of these dynamics, the choice is clear: adapt or risk irrelevance. The business landscape is not a static canvas but a constantly evolving masterpiece. As we embark on our journey through the art of Business Origami, remember that every fold, every twist, and every turn represents an opportunity to adapt, innovate, and thrive in the ever-changing world of business. So, let us proceed, unfold the complexities, and unveil the strategies that will empower you to navigate this dynamic terrain.

Chapter 2: The Art of Business Origami

In the intricate tapestry of business strategy, envision a concept as multifaceted as origami—a strategic framework where adaptability and transformation take center stage. Much like a skilled origamist, successful business leaders wield the principles of Business Origami with vision, precision, and an unwavering commitment to craft something extraordinary. In this chapter, we delve deep into the heart of Business Origami, unfurling its layers to understand the profound intricacies that drive innovation, agility, and transformation in the dynamic world of commerce.

The Essence of Business Origami

At its core, Business Origami is a metaphorical framework that draws inspiration from the art of origami. Origami, the Japanese art of paper folding, is a testament to the transformative power of simplicity and precision. It takes a flat, unassuming sheet of paper and, through a series of meticulously executed folds, transforms it into intricate, three-dimensional creations.

Similarly, Business Origami applies the principles of simplicity, adaptability, and precision to the business world. It acknowledges that in today's fast-paced and ever-evolving environment, businesses must be as nimble as origami creations, ready to pivot and adapt to changing circumstances. Let's explore the key facets that define Business Origami:

Simplicity as a Starting Point

Much like the blank sheet of paper in origami, Business Origami begins with simplicity as its starting point. It encourages businesses to strip away unnecessary complexities and distill their strategies and processes to their core essence. This simplicity serves as the foundation upon which transformative ideas can be built.

In the digital age, where information overload and complexity abound, simplicity is a powerful asset. It allows for clarity of purpose and a streamlined approach to problem-solving. Successful businesses understand that simplicity doesn't equate to lack of sophistication; rather, it is the elegance of a well-crafted solution.

Precision in Execution

Origami demands precision in every fold, and so does Business Origami in its execution. Precision in decision-making, resource allocation, and strategy implementation is essential. Every fold in the business landscape must be deliberate and well-calculated, just like the folds in an origami masterpiece.

Business leaders must be keenly aware of their objectives and the intricate interplay of factors that influence success. Precision ensures that resources are utilized efficiently, and strategies are executed with a high degree of accuracy. It minimizes waste and maximizes the impact of each action.

Adaptability as a Guiding Principle
One of the most vital principles of Business Origami is adaptability. Just as an origamist adjusts their folds in response to the paper's shape and characteristics, businesses must be adaptable to the evolving marketplace. This adaptability is not a reactionary response but a proactive strategy.

The business landscape is ever-changing, influenced by technological advancements, shifting consumer preferences, and global events. To thrive in this environment, companies must be willing to pivot, innovate, and embrace change. Adaptability is the means by which businesses stay relevant and resilient in the face of uncertainty.

Transformation as the End Goal
The ultimate objective of Business Origami is transformation. Origami takes a flat sheet of paper and transforms it into something entirely new and unexpected. Similarly, Business Origami aims to transform businesses by unlocking their hidden potential and enabling them to achieve extraordinary outcomes.

Transformation in Business Origami goes beyond incremental improvements; it involves a radical shift in mindset and approach. It's about reimagining business models, challenging the status quo, and pushing the boundaries of what's possible. Successful transformations are not mere facelifts but profound reinventions that position businesses for long-term success.

The Business Origami Process

Now that we've defined the essence of Business Origami, let's delve into the process by which this metaphorical framework unfolds in the business world. Just as origami has a step-by-step process for creating intricate designs, Business Origami follows a structured approach to achieving adaptability and transformation:

Assess the Starting Point
Every origami creation begins with an assessment of the starting point—the flat sheet of paper. Similarly, businesses must conduct a thorough assessment of their current state. This involves a candid evaluation of strengths, weaknesses, opportunities, and threats (SWOT analysis).
Businesses need to understand where they stand in the market, what their core competencies are, and where there is room for improvement. This assessment serves as the foundation upon which Business Origami strategies are built.

Define the Vision
In origami, the artist visualizes the final creation before making the first fold. Likewise, Business Origami requires a clear vision of the desired outcome. This vision encompasses not only financial goals but also the broader impact a business seeks to achieve.
A well-defined vision serves as a guiding star, aligning all efforts toward a common purpose. It inspires and motivates teams, setting the stage for meaningful transformation.

Simplify and Streamline

Following the principle of simplicity, Business Origami encourages businesses to simplify and streamline their operations. This involves identifying processes, products, or services that may be overly complex or redundant and streamlining them for greater efficiency.

Simplification reduces overhead, increases agility, and enhances the customer experience. It allows businesses to focus on what truly matters and eliminate distractions.

Precision in Strategy

Just as each origami fold requires precision, Business Origami demands precision in strategy development and execution. This includes setting clear objectives, identifying key performance indicators (KPIs), and developing strategies that are aligned with the overall vision.

Precision in strategy ensures that every action taken contributes to the desired transformation. It minimizes the risk of misallocation of resources and maximizes the likelihood of success.

Embrace Adaptability

Adaptability is at the core of Business Origami. Businesses must be prepared to adapt to changing market dynamics, customer preferences, and technological advancements. This requires a culture of continuous learning and innovation.

Adaptability also involves anticipating potential challenges and having contingency plans in place. It's about being agile and responsive to both opportunities and threats.

6. Foster a Culture of Transformation

Transformation is not a one-time event but an ongoing process. Business Origami encourages businesses to foster a culture of transformation, where employees are empowered to contribute ideas and embrace change. This cultural shift involves promoting creativity, collaboration, and a growth mindset. It's about breaking down silos and encouraging cross-functional teams to work together toward a common transformative goal.

The Business Origami Mindset

Beyond the process, Business Origami is a mindset—a way of thinking that permeates every aspect of an organization. Here are the key elements of the Business Origami mindset:

Openness to Change

Business Origami begins with an openness to change. It's about acknowledging that the business landscape is dynamic, and embracing change is essential for survival and growth.

Continuous Learning

Origami artists continually refine their skills, and similarly, businesses must prioritize continuous learning. This includes staying updated on industry trends, exploring new technologies, and seeking inspiration from diverse sources.

Collaboration

Just as origami can be a collaborative art, Business Origami thrives on collaboration. It encourages teams to work together, share ideas, and co-create transformative solutions.

Resilience
Origami creations are resilient, able to withstand challenges and retain their form. Businesses with a Business. Origami mindset build resilience by adapting to adversity and emerging stronger.

Vision-Driven
A clear vision is the North Star of Business Origami. It guides decision-making and inspires innovation. A shared vision unites teams and fuels transformative efforts.

Real-World Examples of Business Origami

To illustrate the principles of Business Origami in action, let's explore real-world examples of companies that have applied this metaphorical framework to achieve adaptability and transformation:

Apple Inc.
Apple's journey from a computer manufacturer to a global technology and entertainment powerhouse exemplifies Business Origami in action. Apple's simplicity-focused approach, precision in design and marketing, and adaptability to changing consumer demands have been key to its success. The company's transformative vision, under

the leadership of Steve Jobs, reshaped industries and set new standards for innovation.

Amazon

Amazon's evolution from an online bookstore to the world's largest e-commerce and cloud computing company is a testament to Business Origami principles. Amazon's relentless focus on customer-centricity, precision in logistics and supply chain management, and adaptability in diversifying its offerings have propelled it to a position of global dominance.

Netflix

Netflix's transformation from a DVD rental service to a streaming media giant is a prime example of Business Origami in the entertainment industry. The company's vision of disrupting traditional television, combined with a commitment to simplifying content consumption and adapting to new technologies, has reshaped how the world watches TV.

Tesla

Tesla's journey to redefine the automotive industry through electric vehicles and sustainable energy solutions embodies the spirit of Business Origami. The company's visionary approach, precision in electric vehicle design, and adaptability in scaling production and expanding its product portfolio have revolutionized the automotive sector.

The Promise of Business Origami

In the world of business, where change is constant and uncertainty prevails, Business Origami offers a guiding philosophy that transcends traditional strategies. It empowers businesses to navigate the complexities of the modern landscape with simplicity, precision, adaptability, and a transformative mindset.

As we embark on this exploration of Business Origami, remember that it is not a one-size-fits-all solution but a framework that can be tailored to each business's unique challenges and opportunities. Just as origami artists craft diverse creations from a single sheet of paper, businesses can draw from the principles of Business Origami to fold their own paths to success.

In the chapters that follow, we will delve deeper into the practical applications of Business Origami, exploring how businesses of all sizes and industries can leverage its principles to achieve adaptability, innovation, and enduring transformation. Our journey has just begun, and the possibilities are as limitless as the folds of an origami masterpiece

Part II: The Seven Folds of Success

In this captivating segment of our journey, we dive deep into the heart of Business Origami's seven essential folds. Just as a master origamist crafts intricate designs through precise folds, these chapters unveil the strategic layers that shape your path to success. From visionary leadership to organizational agility, each fold embodies a transformative principle that propels your business toward resilience and innovation.

Chapter 3: The First Fold - Visionary Leadership

Unfolding Leadership Paradigms

In the ever-evolving landscape of business, leadership stands as the keystone—the first fold in the intricate art of Business Origami. It is a fold that holds the promise of shaping the organization's destiny, just as the initial crease in a sheet of paper defines the course of an origami masterpiece. In this chapter, we embark on a journey into the realm of visionary leadership, where conventional paradigms are unfolded and reimagined. As we explore the essential principles of this first fold, prepare to be captivated by the transformational power of leadership that inspires and propels businesses toward unprecedented heights.

Rethinking Leadership in the Digital Age
The business world today is a far cry from what it was even a decade ago. The digital revolution has redrawn the boundaries of industry, commerce, and innovation. In this dynamic era, leadership paradigms must evolve to keep pace with the rapid changes. Visionary leadership emerges as the cornerstone—a fold that sets the stage for Business Origami's transformative journey.

The Traditional Paradigm
Traditionally, leadership was often associated with a top-down approach—a hierarchical structure where authority

flowed from the top to the bottom. Leaders were expected to possess a set of defined skills, exercise control, and provide direction to subordinates. While this model had its merits, it often lacked the agility needed to thrive in a rapidly changing environment.

The Visionary Paradigm

Visionary leadership redefines the traditional paradigm. It centers on the power of inspiration and purpose. Rather than relying solely on authority, visionary leaders inspire their teams with a compelling vision—an image of a better future that resonates with the organization's values and mission.

Visionary leaders possess the ability to anticipate industry trends, envision opportunities, and craft a path toward their realization. They are architects of the future, capable of folding the organization's potential into innovative strategies. Their approach is not about control but about empowerment, collaboration, and shared ownership of the vision.

Crafting a Vision that Inspires

The heart of visionary leadership lies in crafting a vision that inspires, ignites passion, and propels the organization forward. But how does one create such a vision? Let's explore the art of vision-crafting, uncovering innovative approaches that captivate and rally teams.

Purpose-Driven Vision

A visionary leader's vision is anchored in purpose. It goes beyond profit margins and market share—it speaks to the deeper meaning of the organization's existence. Purpose-driven visions resonate with employees, customers, and stakeholders on a fundamental level.

The Origami Approach: Elevate your vision by asking "why." Explore the core purpose of your organization, beyond financial success. Define the positive impact it seeks to make in the world.

In the heart of Silicon Valley, a visionary leader emerged who would forever redefine the landscape of personal computing. Steve Jobs, co-founder of Apple Inc., was a master of purpose-driven vision. His vision was not just about creating cutting-edge technology; it was about empowering individuals to "think different" and change the world. His famous "1984" commercial, launching the Macintosh, wasn't about product features; it was a call to challenge the status quo and break free from conformity. Steve Jobs's approach to crafting a purpose-driven vision involved asking profound questions: "Why does Apple exist? What do we stand for? What impact can we have on the world?" His answers to these questions shaped Apple's vision and its commitment to innovation, design excellence, and pushing the boundaries of technology.

Inclusivity and Diversity

In an era that values inclusivity and diversity, visionary leaders recognize the importance of involving diverse perspectives in vision-crafting. A vision that reflects the richness of different backgrounds and experiences is more likely to resonate with a broader audience.

The Origami Approach: Create forums for diverse voices to contribute to the vision. Encourage open dialogue and celebrate a variety of viewpoints.

One of the most celebrated examples of inclusivity in vision-crafting comes from the world of literature and entertainment. J.K. Rowling, the author of the Harry Potter series, didn't just craft a story; she created an entire magical world. What sets her vision apart is the inclusivity of characters and themes.

Rowling's Hogwarts School of Witchcraft and Wizardry is a place where diversity thrives. It's not just about the chosen one, Harry Potter; it's about the collective strength of a diverse group of characters—Hermione Granger, Ron Weasley, Hagrid, Luna Lovegood, and many others. Her vision celebrates differences and underscores the idea that everyone has a role to play in shaping a better world. This inclusivity resonated with readers of all backgrounds and ages, making the Harry Potter series a global phenomenon.

Foresight and Adaptability
A visionary leader's vision isn't static; it's a living, adaptable concept. It anticipates changes in the business landscape and incorporates flexibility. It's a vision that acknowledges the uncertainty of the future and embraces it as an opportunity for growth.

The Origami Approach: Develop scenarios that explore potential future scenarios. Plan for contingencies and ensure your vision can adapt to evolving circumstances.

Amazon's Jeff Bezos embodies the essence of foresight and adaptability in vision-crafting. When Bezos founded Amazon in 1994 as an online bookstore, his vision wasn't confined to selling books. He foresaw the potential of e-commerce and, more importantly, the power of customer-centricity.

Bezos's vision was adaptive from the beginning. He recognized that e-commerce would evolve, and he positioned Amazon to adapt with it. The company expanded into diverse product categories, pioneered innovations like one-click shopping and Prime membership, and invested heavily in logistics and fulfillment centers.

But Bezos's foresight extended beyond e-commerce. He envisioned Amazon as a technology powerhouse, with ventures into cloud computing (Amazon Web Services) and digital streaming (Amazon Prime Video). He understood the importance of adaptability, ensuring that Amazon's vision was not constrained by a singular definition of the business.

Storytelling
Visionary leaders are masterful storytellers. They craft narratives that breathe life into the vision, making it

relatable and emotionally engaging. Stories create a sense of purpose and connect individuals to the larger mission.

The Origami Approach: Develop a compelling narrative around your vision. Use storytelling techniques to make it vivid, relatable, and emotionally resonant.

One of the most captivating storytellers in the world of technology is Elon Musk, the visionary behind SpaceX. Musk's vision of making humanity a multi-planetary species is not just a scientific ambition; it's a grand narrative that sparks the imagination.

Musk doesn't simply talk about rocket launches and Mars colonization; he weaves a story of human exploration, survival, and the quest for the unknown. His vision is not just about the technical details of rocketry; it's about the human spirit reaching for the stars.

In Musk's storytelling, the audience becomes part of the adventure. Whether he's discussing the challenges of interplanetary travel or the importance of sustainability, he paints a vivid picture that invites people to join the journey. His storytelling transforms complex concepts into relatable and emotionally charged narratives, rallying individuals to support his vision.

Empowerment
Empowerment is a fundamental tenet of visionary leadership. Leaders empower their teams to contribute to

the vision's realization. This sense of ownership fosters commitment and dedication.

The Origami Approach: Delegate responsibility and provide the autonomy for team members to make decisions that align with the vision

Empowerment fosters innovation and initiative. A shining example of empowerment in action comes from the tech giant Google. Larry Page and Sergey Brin, Google's co-founders, built a culture of empowerment that encouraged employees to explore their innovative ideas.

Google's "20% time" policy allowed employees to dedicate a portion of their work hours to pursuing projects of personal interest. This policy resulted in the creation of some of Google's most significant products, including Gmail and Google News. The company's commitment to employee empowerment led to a culture of innovation, where everyone felt a sense of ownership in contributing to Google's vision of organizing the world's information and making it universally accessible and useful.

Measurable Goals
While a vision is aspirational, visionary leaders understand the importance of grounding it in measurable goals. Clear objectives provide a roadmap for progress and ensure accountability.

The Origami Approach: Break down the vision into concrete, measurable goals and key performance indicators

(KPIs). This allows for progress tracking and adjustments when necessary.

One of the most noteworthy examples of setting measurable goals within a visionary vision is from the world of electric vehicles. Tesla, under the leadership of Elon Musk, had a vision to accelerate the world's transition to sustainable energy. However, this vision wasn't vague; it was accompanied by clear, measurable goals.

Tesla set out to produce electric vehicles that were not just environmentally friendly but also technologically advanced and appealing to consumers. Musk aimed to achieve production volumes that would have a substantial impact on reducing global carbon emissions.

Tesla's vision included concrete targets, such as producing a certain number of electric vehicles per year and expanding the network of Supercharger stations for convenient long-distance travel. These measurable goals ensured that the vision remained on track, and progress was evident to both internal teams and external stakeholders.

Continuous Communication

A visionary leader doesn't simply unveil the vision once and expect it to take root. Instead, they engage in continuous communication, reinforcing the vision's importance and progress.

The Origami Approach: Establish a cadence of communication that keeps the vision at the forefront. Use

multiple channels to reach employees and stakeholders,
and encourage feedback and questions.

One of the most remarkable examples of continuous communication in the realm of vision-crafting comes from the healthcare industry. Dr. Paul Farmer, co-founder of Partners In Health, had a vision of providing quality healthcare to impoverished communities worldwide. He understood that this vision could only succeed if it resonated deeply with the organization's healthcare workers and the communities they served.

Dr. Farmer's approach involved regular, open, and empathetic communication. He made it a point to visit healthcare facilities in underserved areas and engage directly with the staff and patients. He listened to their concerns, understood their needs, and continuously reinforced the vision of providing "a preferential option for the poor in health care." This unwavering communication built trust and commitment among the team, ensuring that the vision was not just a lofty goal but a daily mission.

Real-World Examples of Visionary Leadership

To illustrate the transformative power of visionary leadership, let's turn to real-world examples of leaders who have embraced this paradigm:

Elon Musk - SpaceX
Elon Musk's vision of colonizing Mars is a testament to visionary leadership. He has inspired a dedicated team at

SpaceX to work toward a future where humanity becomes multi-planetary. Musk's vision transcends profit—it speaks to the grandeur of exploration and the survival of our species.

Musk's approach to crafting a visionary mission involves pushing the boundaries of human achievement. SpaceX's mission, "to enable humans to become a multi-planetary species," is not just a tagline; it's a rallying cry that has mobilized engineers, scientists, and dreamers worldwide. Musk's vision goes beyond the technicalities of rocketry; it's a call to humanity to reach for the stars and ensure our species' survival.

Howard Schultz - Starbucks
Howard Schultz's vision for Starbucks extended beyond coffee. He envisioned Starbucks stores as "the third place" between work and home—a gathering place that fosters community. His vision transformed coffee shops into social hubs, and Starbucks into a global brand.

Schultz's approach to visionary leadership involved recognizing the potential for Starbucks to be more than just a coffee retailer. He understood that people seek connection and belonging, and he crafted a vision that resonated with this fundamental human need. Starbucks's commitment to being a place where people could relax, connect, and savor quality coffee fueled its global success.

Indra Nooyi - PepsiCo

Indra Nooyi's tenure as the CEO of PepsiCo was marked by a visionary focus on sustainability. She redefined the company's mission to include "Performance with Purpose," aligning business success with environmental and social responsibility. Her vision transformed PepsiCo into a leader in sustainable practices.

Nooyi's approach to visionary leadership involved challenging the traditional notion of corporate success. She believed that businesses could do well by doing good. Her vision extended beyond profit margins to include positive contributions to society and the environment. PepsiCo's commitment to reducing its environmental footprint, promoting diversity and inclusion, and investing in healthier products exemplified Nooyi's vision of "Performance with Purpose."

The Unfolding Journey

As we conclude this chapter on visionary leadership, we embark on an unfolding journey of transformation. Visionary leadership serves as the first fold, setting the stage for the remaining folds of Business Origami. It is the spark that ignites innovation, aligns teams, and propels organizations toward a future filled with promise and purpose.

In the chapters that follow, we will continue our exploration of Business Origami's folds, each revealing a unique facet of success in the modern business landscape. The journey has just begun, and the possibilities are as boundless as the visionary leaders who inspire us to reach for the stars. As you contemplate the principles of visionary leadership, consider how they can guide your own journey as a leader, and how they can shape the destiny of your organization.

Chapter 4: The Second Fold - Strategic Agility

Adapting to Change

Navigating Uncertainty with Strategy
In the ever-shifting landscape of business, the second fold of Business Origami unfurls before us—Strategic Agility. Just as a master origamist pivots and adjusts with each fold, organizations must cultivate the ability to adapt swiftly and strategically. This chapter explores the dynamic interplay between strategy and agility, revealing how businesses can navigate uncertainty and harness change as a catalyst for growth. As we delve into the art of the second fold, prepare to embark on a journey that uncovers innovative approaches to strategic agility.

The Imperative of Strategic Agility
In today's business world, change is not an occasional visitor; it's a constant companion. Technological advancements, market disruptions, and unforeseen global events can reshape industries overnight. To thrive in such an environment, organizations must embrace strategic agility—a mindset that combines a robust strategy with the flexibility to adapt to evolving circumstances.

The Traditional Approach
Historically, businesses relied on rigid, long-term strategic plans. These plans, often extending several years into the future, provided a roadmap based on assumptions about

stable market conditions. While such plans had their merits, they were ill-suited to the realities of a volatile, interconnected global economy.

The traditional approach to strategy assumed that once a plan was set in motion, it should remain largely unchanged. Deviations from the plan were viewed as disruptions or failures. This approach could lead to businesses becoming locked into strategies that were no longer relevant.

The Agility Imperative

Strategic agility represents a paradigm shift—an evolution from the traditional approach. It acknowledges that change is not a deviation but an inherent part of the business landscape. To thrive, organizations must become adept at sensing change, responding effectively, and even proactively shaping their future.

The Agile Mindset

At the heart of strategic agility lies an agile mindset. This mindset embraces change as an opportunity rather than a threat. It thrives on uncertainty and complexity, viewing them as fertile ground for innovation and growth.

Leaders with an agile mindset recognize that the ability to pivot quickly, experiment with new ideas, and learn from failures is a competitive advantage. They understand that agility is not the absence of a plan but the ability to adapt the plan in real-time.

The Agile Organization

Strategic agility extends beyond individual mindset; it permeates the entire organization. An agile organization is one that can swiftly adjust its strategies, structures, processes, and culture to meet changing circumstances.

Key Elements of an Agile Organization

Adaptive Strategy: Agile organizations embrace iterative strategy development. They continuously assess their strategic direction, adapting to market feedback and emerging trends.

Cross-Functional Collaboration: Silos are dismantled in agile organizations. Cross-functional teams collaborate seamlessly to respond to challenges and opportunities.

Rapid Decision-Making: Agile organizations empower teams at all levels to make decisions swiftly. This reduces bureaucracy and accelerates execution.

Customer-Centricity: Customer feedback is a central driver of agile strategies. Organizations prioritize understanding and meeting customer needs.

Experimentation and Learning: Agile organizations encourage experimentation and embrace failure as a source of learning. They foster a culture of continuous improvement.

Approaches to Strategic Agility

Embracing strategic agility requires a multifaceted approach. Let's explore innovative strategies and approaches that can enhance an organization's agility:

Scenario Planning

Scenario planning is a proactive approach to strategic agility. It involves developing multiple scenarios or potential futures and planning responses for each. By envisioning various possibilities, organizations can better prepare for uncertainties.

The Origami Approach: Develop a range of scenarios that encompass different economic, political, and technological conditions. Assign teams to each scenario to formulate strategies and action plans.

Agile Product Development

Innovation and product development are at the heart of many agile organizations. They prioritize delivering value to customers quickly through iterative product development cycles.

The Origami Approach: Implement agile methodologies such as Scrum or Kanban to break down projects into smaller, manageable tasks. Regularly gather feedback from customers to inform product iterations.

Open Innovation

Open innovation involves collaborating with external partners, including customers, suppliers, and even competitors, to drive innovation and expand capabilities.

The Origami Approach: Establish partnerships and open channels for idea exchange with external stakeholders. Leverage external expertise and resources to accelerate innovation.

Design Thinking

Design thinking is a human-centered approach to problem-solving that encourages empathy, creativity, and iteration. It can be a powerful tool for agile organizations seeking to address complex challenges.

The Origami Approach: Implement design thinking workshops and processes to tackle strategic problems. Encourage cross-functional teams to empathize with end-users, ideate, prototype, and test solutions.

Lean Startup Methodology
The Lean Startup methodology emphasizes experimentation, validated learning, and iterative product development. It's particularly effective for startups and established organizations looking to launch new ventures.

The Origami Approach: Apply Lean Startup principles such as building a minimum viable product (MVP), measuring its performance, and iterating based on feedback. This approach reduces the risk of launching a product that doesn't resonate with customers.

Real-World Examples of Strategic Agility

To illustrate the concept of strategic agility in action, let's explore real-world examples of organizations that have successfully navigated change through agile strategies:

Toyota - Lean Production

Toyota's journey to becoming a leader in the automotive industry is synonymous with strategic agility. The company pioneered the lean production system, which focused on reducing waste and improving efficiency.

Toyota's approach involved empowering employees to identify and solve production issues, implementing just-in-time inventory systems, and maintaining a relentless commitment to continuous improvement. This agility allowed Toyota to adapt to market fluctuations, changing consumer demands, and global economic challenges.

Spotify - Revolutionizing Music Streaming

Spotify's agile approach involved developing a user-friendly platform with both free and premium subscription options. The company actively engaged with artists and labels to secure a vast catalog of music. Continuous algorithm improvements personalized user experiences, making music discovery seamless.

In response to industry shifts and competitors, Spotify expanded into podcasts, making strategic acquisitions like Gimlet Media and Anchor. This pivot positioned Spotify as a leading player in the audio content space. Spotify's ability to anticipate trends, pivot swiftly, and create a platform that resonated with users exemplifies strategic agility.

Zoom Video Communications - Redefining Virtual Communication

Zoom's strategic agility involved rapidly developing and refining its video conferencing platform. The company prioritized ease of use, crystal-clear audio, and stable video connections. During the COVID-19 pandemic, Zoom became a household name, demonstrating its ability to scale and meet increased demand.

Zoom also embraced innovation by continuously introducing new features, such as breakout rooms, virtual backgrounds, and integrations with productivity tools. These innovations allowed Zoom to serve a diverse range of users, from businesses to educational institutions and individuals. Zoom's responsiveness to emerging needs and its commitment to improving the virtual meeting experience highlight its strategic agility.

Navigating Uncertainty with Strategy

Strategic agility is not about abandoning long-term goals or plans; it's about ensuring that those goals remain relevant in an ever-changing world. As we conclude this exploration of the second fold, remember that strategic agility is an unfolding journey—a dynamic interplay between strategy and adaptability.

In the chapters that follow, we will continue to unravel the folds of Business Origami, each revealing a unique facet of success in today's business landscape. As you reflect on the principles of strategic agility, consider how your organization can embrace change as an opportunity, and how you can lead the way in navigating uncertainty with a strategic mindset that thrives on adaptation and innovation.

Chapter 5: The Third Fold - Innovative Thinking

The Origami Mindset

Fostering Creativity and Innovation

In the intricate art of Business Origami, the third fold unfurls before us—the realm of Innovative Thinking. Just as an origamist transforms a simple sheet of paper into a masterpiece through imagination and dexterity, organizations can harness the power of innovative thinking to craft unparalleled solutions and drive sustainable success. This chapter transcends conventional business concepts, inviting you to embark on a journey where creativity and innovation reign supreme.

The Essence of Innovative Thinking

Innovation is the lifeblood of progress, and it thrives in an environment where imaginative thinking is encouraged. While innovation isn't a novel concept in the business world, the Origami Mindset we're about to explore transcends traditional approaches, offering a fresh perspective on fostering creativity and innovation within organizations.

Beyond Incremental Improvements

Many organizations limit their understanding of innovation to incremental improvements—small enhancements to existing products, services, or processes. While these incremental innovations are undoubtedly valuable, the

Origami Mindset challenges us to think beyond these confines.

The Origami Approach: Encourage teams to view innovation as a quest for transformative breakthroughs, akin to unfolding an entirely new dimension within the business landscape. Challenge the status quo, ask audacious questions, and pursue innovative ideas that transcend the incremental.

Key Principles of The Origami Mindset

At its core, the Origami Mindset is a philosophy that embraces unconventional thinking, creativity, and a willingness to deconstruct established norms. Just as an origamist sees limitless potential in a simple sheet of paper, individuals and organizations adopting this mindset recognize the boundless possibilities within their grasp.

Curiosity Unleashed: The Origami Mindset encourages relentless curiosity, urging individuals to explore the unknown, question assumptions, and seek inspiration from diverse sources.

Fearless Imagination: To fold new origami creations, one must let go of fear and embrace the unknown. Likewise, the Origami Mindset inspires fearless imagination, inviting individuals to envision possibilities without constraints.

Iterative Experimentation: Just as an origamist iteratively refines their creation, the Origami Mindset promotes experimentation and learning from failures. It recognizes that innovation often emerges from the willingness to iterate and adapt.

Embrace Complexity: Complexity is not a hurdle but a canvas for innovation. The Origami Mindset encourages individuals to tackle intricate challenges, knowing that the most profound solutions often emerge from the depths of complexity.

Unconventional Collaborations: Origami's elegance unfolds when different folds come together. Similarly, the Origami Mindset champions unconventional collaborations, bringing together individuals with diverse perspectives, skills, and backgrounds to spark innovative ideas.

Approaches to Fostering Innovative Thinking

The Origami Mindset extends beyond philosophy; it manifests in tangible approaches that organizations can adopt to nurture innovative thinking. Let's explore these approaches, each designed to challenge conventional boundaries and foster a culture of boundless creativity.

Moonshot Thinking
Originating from Google's X division, Moonshot Thinking is a concept that encourages organizations to set audacious, game-changing goals. Rather than aiming for incremental improvements, Moonshot Thinking pushes teams to

envision solutions that are ten times better, fundamentally transformative, and previously thought impossible.

The Origami Approach: Embrace Moonshot Thinking as the standard for goal-setting. Challenge teams to identify moonshot opportunities within their domains. Encourage audacious problem-solving by asking, "What if there were no limits?".

Innovation Ecosystems

Innovation thrives in diverse ecosystems where ideas can cross-pollinate. Building innovation ecosystems involves forging partnerships, both internal and external, to create a fertile ground for innovative thinking.

The Origami Approach: Cultivate innovation ecosystems that extend beyond traditional boundaries. Collaborate with startups, universities, and unconventional partners to bring fresh perspectives and ideas into your organization. Foster a culture of open innovation where the flow of ideas is unrestricted.

Design Thinking Beyond Products

While Design Thinking is typically associated with product development, the Origami Mindset extends this approach beyond products. Design Thinking can be applied to processes, customer experiences, and even organizational structures.

The Origami Approach: Encourage teams to use Design Thinking principles to reimagine every aspect of the organization. From the employee onboarding experience to supply chain management, the Origami Mindset challenges individuals to design for innovation in every corner of the business.

Creativity Sparks

Innovation often begins with a spark of creativity. The Origami Approach encourages individuals to seek inspiration from unconventional sources and experiences.

The Origami Approach: Organize "Creativity Sparks" sessions where teams engage in activities or experiences outside their usual domains. These sessions can include art exhibitions, nature excursions, or brainstorming sessions in non-traditional settings. Encourage participants to draw connections between these experiences and their work.

Inclusivity and Diversity

Innovative thinking thrives in diverse and inclusive environments where different perspectives are valued. The Origami Mindset recognizes the power of diverse teams in generating creative solutions.

The Origami Approach: Actively promote diversity and inclusivity in all aspects of the organization. Create platforms for underrepresented voices to be heard and valued. Ensure that decision-making processes incorporate a wide range of perspectives.

Real-World Examples of The Origami Mindset

To illustrate the transformative power of the Origami Mindset, let's delve into real-world examples of organizations that have embraced unconventional thinking to drive innovation:

Ginkgo Bioworks - Bioengineering for the Future
Ginkgo Bioworks, a biotechnology company, redefines the possibilities of bioengineering. The company's Origami Approach involves leveraging biology as a platform for innovation. Ginkgo Bioworks engineers microorganisms to produce valuable materials and chemicals, disrupting traditional manufacturing processes.
Ginkgo Bioworks' innovative thinking extends to diverse applications, from producing fragrances using yeast to engineering microbes for sustainable agriculture. The company's commitment to bioengineering innovation challenges conventional approaches to production, offering a glimpse into a future where biology is a powerful tool for sustainable and efficient manufacturing.

Nintendo - Redefining Gaming
Nintendo, a pioneer in the gaming industry, has consistently demonstrated innovative thinking. Rather than competing solely on hardware specifications, Nintendo focuses on creating unique gaming experiences that captivate players of all ages.
Nintendo's Origami Approach involves designing consoles like the Nintendo Switch, which seamlessly transitions between handheld and home gaming. This innovation

offers flexibility and engages players in novel ways. Games like "Wii Sports" and "Animal Crossing" challenge traditional gaming concepts, emphasizing interactive and social gameplay.

Pixar - Creative Storytelling

Pixar Animation Studios, renowned for its animated films, epitomizes innovative thinking in storytelling. Beyond technological advancements, Pixar's success lies in its ability to craft emotionally resonant narratives that appeal to audiences of all ages.

Pixar's Origami Approach to storytelling involves unconventional thinking, exploring deep and universal themes, and challenging storytelling conventions. The studio encourages creative teams to embrace vulnerability and push the boundaries of storytelling. The result is a portfolio of films that transcend traditional animation, leaving a lasting impact on the art of cinematic storytelling.

Unfolding Possibilities

As we conclude this chapter on Innovative Thinking and the Origami Mindset, we enter a realm where creativity knows no bounds. The Origami Mindset invites individuals and organizations to transcend the ordinary, challenge the status quo, and envision solutions that defy convention.

In the chapters that follow, we will continue our journey through the folds of Business Origami, each unveiling a unique facet of innovation and success. As you explore the principles of the Origami Mindset, consider how your organization can foster a culture where innovative thinking flourishes, and how you can lead the way in unfolding possibilities that redefine your industry and shape the future.

Chapter 6: The Fourth Fold - Adaptive Culture

Building a Culture of Agility

In the intricate folds of Business Origami, we find the fourth fold—the domain of Adaptive Culture. Just as an origamist carefully crafts each fold to create a harmonious masterpiece, organizations can cultivate a culture that adapts seamlessly to change, fostering resilience and innovation. This chapter transcends conventional notions of corporate culture, inviting you to embark on a transformative journey where change is not an obstacle but a constant companion.

The Heart of Adaptive Culture

An adaptive culture is the beating heart of an organization's ability to flourish in today's ever-evolving business landscape. While the concept of organizational culture is not new, the Origami Mindset approach challenges us to reimagine and redefine what culture means in the context of adaptability and change.

Beyond Rigid Structures

Many organizations are built on traditional cultural foundations, emphasizing stability and predictability. However, let's step outside these boundaries and imagine a culture that's as dynamic as the world it operates in.

The Origami Approach: Picture a culture that values flexibility, experimentation, and adaptability. One that not only copes with change but thrives on it. Challenge the idea of a static corporate culture and embrace one that evolves with the dynamic business environment.

The Essence of Adaptive Culture
At its core, the Adaptive Culture is a philosophy that sees change as an opportunity rather than a threat. It's a culture that recognizes the constant evolution of the business landscape and equips organizations with the tools to not just survive but to flourish in times of uncertainty.

Core Principles of the Adaptive Culture

Change as a Constant: In the Adaptive Culture, change is not an anomaly; it's a way of life. It's like the weather, ever-changing and unpredictable. Imagine a culture where change is not feared but expected and embraced.

Inclusive Decision-Making: Adaptive cultures value the input of everyone, from top executives to entry-level employees. Decisions are made collectively, drawing from diverse perspectives and expertise.

Rapid Experimentation: The Adaptive Culture encourages teams to experiment and iterate quickly. Failures aren't viewed as setbacks but as stepping stones to innovation.

Transparency and Communication: Open and transparent communication is paramount. In this culture, employees are informed about changes, the reasons behind them, and how they fit into the bigger organizational picture.

Continuous Learning: Learning isn't an occasional event; it's ingrained in the culture. Employees are encouraged to learn, adapt, and stay ahead of industry trends. Imagine a workplace where growth never stops.

Approaches to Building an Adaptive Culture

But how do we transform these principles into reality? The Adaptive Culture isn't just a philosophy; it's a set of concrete approaches that organizations can adopt to cultivate a culture of agility.

Change Champions

Change doesn't happen in isolation; it needs champions. Champions are the individuals who lead by example, who not only embrace change but inspire others to do the same.

The Origami Approach: Think of these champions as the artists of your organization. Develop a Change Champion Program that recognizes and celebrates employees who demonstrate adaptability and resilience in the face of change. Encourage them to mentor and support their peers in navigating change.

Learning Ecosystem

Imagine an ecosystem where knowledge flows freely, nourishing the minds of your employees. Create a learning ecosystem that empowers them to acquire new skills and knowledge continually.

The Origami Approach: In this ecosystem, learning is not a chore; it's a joy. Establish a digital learning platform that offers a wide range of courses and resources. Encourage employees to allocate a portion of their working hours to learning and development. Recognize and reward employees who demonstrate a commitment to continuous learning.

Agile Teams

Think of your organization as a collection of agile teams that can pivot swiftly in response to changing market dynamics.

The Origami Approach: These teams aren't bound by rigid structures. Implement agile methodologies like Scrum or Kanban in various departments. Encourage cross-functional collaboration and empower teams to make decisions autonomously. Celebrate team achievements and innovations that result from this approach.

Change-Ready Leadership

Leadership isn't about maintaining the status quo; it's about guiding the organization through change.

The Origami Approach: Leaders are like the masters of origami, shaping the organization's destiny. Provide leadership training that focuses on leading through change. Equip leaders with the skills to communicate effectively during transitions and inspire their teams to embrace change. Recognize and promote leaders who excel in leading change initiatives.

Feedback Loops
Imagine a workplace where every employee's voice matters, where feedback flows freely, like a gentle breeze.

The Origami Approach: In this workplace, feedback isn't a one-way street. Implement regular pulse surveys or feedback mechanisms that enable employees to share their thoughts on recent changes and suggest areas for improvement. Act on this feedback to demonstrate a commitment to employee well-being and adaptability.

Real World Examples of the Adaptive Culture

To illustrate the transformative power of the Adaptive Culture, let's delve into real-world examples of organizations that have successfully embraced change as a constant companion:

Microsoft - Cultural Transformation
Microsoft, under the leadership of Satya Nadella, embarked on a cultural transformation journey.
Microsoft's Origami Approach involved adopting a growth mindset, where employees were encouraged to learn from

failures and adapt continuously. The company also embraced a more open and collaborative culture, fostering innovation and agility.

Zappos - Holacracy Experiment
Zappos, the online shoe and clothing retailer, experimented with a radical shift in its organizational structure by adopting Holacracy—a self-management system. Zappos' Origami Approach challenged the traditional hierarchy and empowered employees to make decisions autonomously. While the Holacracy experiment had its challenges, it showcased Zappos' commitment to adaptability and a culture that encourages employees to take ownership of their roles.

Google - Agile Workforce
Google, known for its innovative culture, has taken agility to a new level by creating a flexible and adaptive workforce.

Google's Origami Approach involves initiatives like the "20% Time," where employees can spend a portion of their workweek pursuing projects of personal interest. This approach encourages experimentation and allows employees to contribute to the company's innovation efforts actively.

Unfolding Resilience and Innovation

As we conclude this chapter on Adaptive Culture, we recognize that the ability to embrace change as a constant is the hallmark of organizations that thrive in today's dynamic business environment. The Adaptive Culture invites individuals and organizations to transcend fear, embrace resilience, and cultivate innovation amidst uncertainty.

Consider this not just a chapter but an invitation to reshape your organization's culture. As we continue our journey through the folds of Business Origami, imagine how you can lead the way in building a culture that doesn't just survive change—it thrives on it. Change, when approached with the right mindset, can be your greatest ally on the path to innovation and success.

Chapter 7: The Fifth Fold - Customer-Centricity

In the art of Business Origami, we have journeyed through the folds of visionary leadership, strategic agility, innovative thinking, and adaptive culture. Each fold has revealed a unique facet of business excellence, akin to the intricate creases that transform a simple sheet of paper into a masterpiece. As we embark on the fifth fold of our journey, we find ourselves in the realm of Customer-Centricity—a domain where businesses not only acknowledge the importance of their customers but fold their entire existence around them. In this chapter, we embark on a profound journey that transcends the superficial notions of customer-centricity. We delve into the very essence of what it means to fold around the customer and create experiences that resonate deeply.

The Canvas of Customer-Centricity

Imagine a canvas where the brushstrokes are the experiences of your customers. Picture a masterpiece where every detail, every color, and every stroke is meticulously crafted to evoke delight and satisfaction. This is the essence of Customer-Centricity—the art of folding your business around the customer's needs, desires, and aspirations.

The Customer-Centric Mindset
Customer-Centricity is more than just a buzzword; it's a mindset that permeates every facet of an organization. It's

the belief that the customer isn't merely a transaction; they are the heartbeat of your business.

The Origami Approach: Let's envision a world where every decision, every innovation, and every strategy revolves around the customer. This isn't limited to the customer service department; it's a mindset that should pulse through every department, from product development to marketing, from sales to finance.

The Essence of Customer-Centricity

At its core, Customer-Centricity is the understanding that customers are the North Star guiding your business journey. It's about creating experiences that resonate with them, forging emotional connections, and consistently exceeding their expectations.

Key Principles of Customer-Centricity

Deep Customer Understanding
Imagine knowing your customers so well that you can anticipate their needs before they do. It's about diving deep into customer insights, understanding their pain points, aspirations, and motivations.

Personalization at Scale
In the age of personalization, customers expect tailored experiences. Picture a world where your interactions with customers are like bespoke suits, perfectly fitting their preferences.

Feedback as a Gift
Instead of fearing criticism, Customer-Centric organizations embrace feedback. Imagine feedback as a precious gift that helps you refine your products and services.

Seamless Omnichannel Experiences
In the digital age, customers move across channels effortlessly. Envision a seamless transition from your website to your app to your physical store, with a consistent experience at every touchpoint.

Continuous Improvement
Customer-Centric organizations don't rest on their laurels. They're in a constant state of evolution, striving to enhance customer experiences.

Innovation Through Customer Eyes
In this world, innovation isn't a secretive process hidden behind closed doors; it's a collaborative journey with your customers. Customers become co-creators, actively shaping the products, services, and experiences they desire.

Enduring Trust
Trust forms the bedrock of any meaningful relationship. In the customer-centric world, trust is not a goal; it's a constant. It's the unwavering belief that your organization will always act in the customer's best interest.

Approaches to Customer-Centricity

The path to Customer-Centricity isn't linear; it's a journey of continuous improvement and innovation. Let's explore some approaches that can help you fold your business around the customer:

Customer Journey Mapping

Imagine tracing the steps of your customers as they interact with your business. Customer journey mapping is an exercise that helps you understand every touchpoint, from the first awareness to post-purchase support.

The Origami Approach: Create a comprehensive customer journey map that not only identifies touchpoints but also highlights pain points and moments of delight. Use this map to make informed decisions about where to focus your efforts.

Voice of the Customer (VoC) Programs

In a Customer-Centric world, the customer's voice is a guiding star. VoC programs capture customer feedback, opinions, and suggestions.

The Origami Approach: Establish robust VoC programs that include surveys, feedback forms, and social listening. Encourage customers to share their thoughts and ideas openly. Use this data not just to measure satisfaction but to drive meaningful improvements.

Agile Product Development

Picture a world where product development isn't a secretive process but a collaborative effort with your customers. Agile methodologies involve customers in the product development journey.

The Origami Approach: Implement agile product development practices that allow for rapid iterations and customer feedback integration. Engage customers in beta testing and co-creation. This not only results in products that better meet customer needs but also creates a sense of ownership among customers.

AI-Powered Hyper-Personalization
Personalization has graduated from token gestures to a realm where artificial intelligence (AI) crafts seamless, anticipatory experiences.

The Origami Approach: Picture a future where AI doesn't just automate processes but elevates personalization to an art form. Algorithms, powered by vast datasets, predict customer preferences and behaviors. It's akin to having a personal concierge who knows your customers better than they know themselves. From product recommendations to content curation, AI becomes the guardian of hyper-personalized experiences.

Customer-Centric Leadership
Customer-Centricity begins at the top. Leaders must embody the Customer-Centric mindset and inspire their teams to do the same.

The Origami Approach: Envision a leadership team that actively seeks customer feedback and incorporates it into strategic decisions. Leaders should regularly engage with customers, whether through customer panels, surveys, or direct interactions. This not only keeps leaders connected to the customer but also sets an example for the entire organization.

Transparent Data-Driven Trust
Trust isn't established through promises alone; it thrives on transparency and integrity. Visualize a world where data is not a hidden asset but a shared resource.

The Origami Approach: Organizations in this world embrace a new level of transparency. They grant customers access to their own data, showcasing how it's employed and safeguarded. Blockchain technology ensures data security and gives customers full control over who accesses their information. Trust isn't assumed; it's constructed through data-driven integrity.

Real-World Examples of Customer-Centricity

To illustrate the transformative power of Customer-Centricity, let's explore real-world examples of organizations that have mastered the art of folding their businesses around the customer:

Amazon - Customer Obsession

Amazon, one of the world's largest e-commerce giants, thrives on Customer-Centricity. Their Origami Approach involves an unwavering focus on the customer.

Amazon envisions a world where every customer receives what they desire swiftly and conveniently. This has led to innovations like one-click ordering, personalized recommendations, and an extensive network of distribution centers to ensure speedy delivery.

Disney - Magical Experiences

Disney, renowned for creating magical experiences, has a Customer-Centric culture deeply ingrained in its DNA. Their Origami Approach revolves around exceeding customer expectations at every touchpoint. From meticulously designed theme parks to heartwarming storytelling in movies, Disney immerses customers in enchanting experiences.

Nike - Personalized Product Design

Nike, the global sportswear giant, has embraced customer-centricity through personalized product design. They offer NikeID, a platform that allows customers to customize their sneakers, from colors to materials and even adding personalized text or graphics. NikeID turns customers into co-creators, making the product uniquely theirs. By enabling customers to take an active role in the design process, Nike deepens the emotional connection between customers and their products, fostering brand loyalty and advocacy.

Creating Customer-Centric Experiences

As we conclude this chapter on Customer-Centricity, we invite you to envision a world where every interaction with your business leaves customers not just satisfied but delighted. The art of folding around the customer is a journey of continuous refinement and innovation, where every fold represents an opportunity to create value, forge connections, and build lasting relationships.

Chapter 8: The Sixth Fold - Resilient Operations

As we embark on this next fold in our journey through the intricate art of Business Origami, we enter a dimension critical for survival in today's ever-changing landscape. Just as an origamist anticipates each crease to create a masterpiece, organizations must approach their operations with the same level of precision and foresight. This chapter unveils the sixth fold—a realm of resilient operations that prepares businesses to weather any storm.

The Essence of Resilient Operations

In an era of unprecedented disruptions and unforeseen challenges, resilient operations are not a luxury but a necessity. It's the ability to absorb shocks, adapt swiftly, and emerge stronger from adversity. Resilient operations don't just protect businesses; they propel them toward growth.

Beyond Traditional Efficiency
Traditionally, operational efficiency was the gold standard. However, efficiency alone is no longer sufficient. Resilient operations are not merely about doing things right; they're about doing the right things, even when the unexpected occurs.

The Origami Approach: Picture your operations as a dynamic entity, ready to pivot and adapt to any circumstance. It's not just about minimizing costs; it's about maximizing adaptability. Resilience becomes the cornerstone of your operations, ensuring that your organization can thrive in times of uncertainty.

The Pillars of Resilient Operations

Dynamic Scalability

Resilient operations are not rigid structures but agile systems that can scale up or down as needed. Whether it's sudden market fluctuations, supply chain disruptions, or unforeseen surges in demand, your operations should flex without breaking.

Supply Chain Redundancy

In an interconnected world, supply chains are susceptible to disruptions. Resilient operations involve identifying and mitigating vulnerabilities within your supply chain. This includes diversifying suppliers, reducing dependencies, and building redundancy.

Data-Driven Decision-Making

Resilience is rooted in information. Access to real-time data enables organizations to make informed decisions rapidly. The Origami Approach involves creating a data-centric culture where decisions are driven by insights rather than assumptions.

Cross-Functional Collaboration

Silos inhibit resilience. Cross-functional collaboration ensures that various parts of the organization work in harmony, sharing insights and resources to address challenges collectively.

Risk Anticipation and Mitigation
Resilient operations aren't just about reacting to shocks; they're about anticipating them. This involves comprehensive risk assessment and proactive measures to mitigate potential disruptions.

Approaches to Resilient Operations

But how do we transform these principles into action? How do we fold our operations to be more resilient in a world of uncertainty? Let's delve into unconventional approaches that challenge conventional wisdom and offer new insights into building resilient operations.

Swarm Intelligence
Nature often holds the key to innovation. One remarkable strategy is drawn from the collective intelligence of organisms like bees and ants. These creatures exhibit remarkable coordination without a central authority.

The Origami Approach: Imagine implementing swarm intelligence in your supply chain. Instead of relying solely on a hierarchical structure, your supply chain could be designed to adapt autonomously. Sensors and algorithms allow components to communicate and make real-time decisions based on changing conditions. It's a distributed, self-organizing approach that optimizes operations without centralized control.

Circular Economy
The traditional linear economy is based on take-make-waste. In contrast, the circular economy promotes the reduction, reuse, and recycling of resources, creating a closed-loop system.

The Origami Approach: Embrace the circular economy in your operations. Design products with recyclability in mind. Implement processes that minimize waste and promote the repurposing of materials. Not only does this enhance sustainability, but it also reduces vulnerabilities related to resource scarcity.

Scenario Gaming
Traditional risk assessments often fall short in predicting complex, interconnected disruptions. Scenario gaming involves creating detailed narratives of potential crises and conducting simulated exercises to test organizational responses.

The Origami Approach: Regularly engage in scenario gaming for your operations. Develop scenarios that challenge your organization's adaptability, such as a sudden supply chain disruption or a cyberattack. These exercises not only prepare your teams but also reveal weaknesses in your operations that can be addressed proactively.

Digital Twins

Digital twins are virtual replicas of physical objects, processes, or systems. They provide real-time insights, enabling organizations to monitor, analyze, and optimize operations.

The Origami Approach: Create digital twins of your critical operational components. This allows for predictive maintenance, real-time monitoring, and rapid response to anomalies. It's a level of visibility that enhances resilience by enabling proactive intervention.

Real-World Examples of Resilient Operations

To illustrate the transformative power of resilient operations, let's explore a real-world example:

Walmart - The Supply Chain Innovator
Walmart, one of the world's largest retailers, has invested heavily in building a resilient supply chain. They leverage technology, including blockchain, to track the journey of products from source to shelf. This not only enhances transparency but also enables rapid traceability in case of recalls or disruptions. Additionally, Walmart maintains a private fleet of trucks, giving them greater control over logistics and reducing dependence on external carriers.

Procter & Gamble (P&G) - The Resilience of Predictive Analytics
P&G, a consumer goods giant, utilizes predictive analytics to enhance its supply chain resilience. By analyzing data from various sources, including weather forecasts and

historical sales data, P&G can anticipate disruptions like natural disasters or sudden shifts in consumer demand. This data-driven approach enables them to adjust production and distribution in real-time, minimizing the impact of unexpected events.

Johnson & Johnson - Diversification for Resilience
Pharmaceutical and consumer goods company Johnson & Johnson has demonstrated the importance of diversification to ensure operational resilience. By operating in multiple product categories, including pharmaceuticals, medical devices, and consumer products, J&J has reduced its exposure to disruptions in any single sector. This diversification strategy has enabled the company to maintain stability and adapt to changing market conditions over the years.

The Future of Resilient Operations

As we conclude this chapter on Resilient Operations, I invite you to reflect on the transformative potential of the Origami Approach. Resilience is not a passive state; it's an active pursuit. In a world where uncertainty is the only constant, organizations must fold their operations with precision and foresight. By understanding these approaches you can reshape your organization's operations for a future filled with unforeseen challenges. Just as an origamist anticipates each fold to create a masterpiece, you can anticipate and adapt to business shocks, emerging stronger and more resilient with each challenge you fac

Chapter 9: The Seventh Fold - Sustainable Growth

As we continue our journey through the intricate art of Business Origami, we arrive at the seventh fold—a realm where the pursuit of sustainable growth takes center stage. Just as an origamist carefully balances each fold to craft an enduring masterpiece, organizations must tread the delicate path between expansion and stability to ensure their long-term prosperity.

Sustaining Success Over Time

Sustainable growth isn't just about achieving momentary success; it's about creating a legacy that endures through time. It's a harmonious dance between expansion and stability, innovation and tradition, progress and preservation. This chapter explores how organizations can foster sustainable growth by drawing lessons from history and gazing into the future.

Beyond Short-Term Triumphs

In an era where quarterly reports often dictate strategies, sustainable growth challenges the allure of immediate gains. It prompts organizations to shift their perspective from short-term triumphs to long-term impact.

The Origami Approach: Picture your organization as a steward of the future, responsible not only for maximizing shareholder value but also for preserving the planet, nurturing communities, and safeguarding resources for generations to come. It's a perspective that transcends profits alone and places equal importance on purpose and impact.

The Historical Perspective

To understand sustainable growth, we must first look to history. The concept of sustainable growth isn't a modern invention but an age-old practice of civilizations and empires that recognized the importance of balance. Throughout the ages, civilizations, empires, and businesses have risen and fallen. What separates those that endure from those that fade into obscurity?

Lessons from Ancient Rome

The Roman Empire provides a poignant lesson in the perils of unsustainable growth. Driven by expansion and conquest, it grew to immense proportions. Yet, its relentless quest for new territories and resources ultimately led to overextension, resource depletion, and social unrest. Rome serves as a stark reminder that unchecked growth can lead to collapse.

The Origami Approach: Organizations must learn from Rome's mistakes by adopting a holistic perspective. Growth should not be pursued at the expense of long-term sustainability. Just as Rome could have benefited from prudent resource management, modern organizations must prioritize sustainable practices that ensure they thrive for centuries, not just years.

The Resilience of Family Businesses

Family businesses often exemplify the principles of sustainable growth. Passed down through generations, these enterprises prioritize continuity and resilience over rapid expansion. The commitment to preserving the family legacy promotes a long-term perspective.

The Origami Approach: Even in non-family organizations, adopting a family business mindset can be transformative. Prioritizing stability, values, and a multi-generational outlook fosters sustainable growth by ensuring that decisions made today consider their impact on the organization's future.

The Pillars of Sustainable Growth

Sustainable growth relies on several fundamental pillars, each contributing to the enduring success of an organization:

Purpose-Driven Innovation: Sustainable growth thrives on innovation that aligns with a higher purpose. It's about creating products and services that not only meet market demands but also address societal and environmental challenges.

Triple Bottom Line: Traditional financial metrics give way to a triple bottom line that considers people, planet, and profits. Organizations measure success not only in terms of economic gains but also through social and environmental impact.

Long-Term Perspective: Sustainable growth requires a shift from short-term thinking to long-term planning. Organizations must make decisions today that benefit not just the current generation but those to come.

Stakeholder Engagement: Sustainable growth embraces a broader array of stakeholders, including employees, customers, communities, and the natural world. It's about fostering partnerships and shared responsibility.

Approaches to Sustainable Growth

But how do we breathe life into these principles? How do we fold our organizations to foster sustainable growth in a world defined by rapid change and disruption? Let's explore the approaches that challenge the conventional wisdom of endless growth.

Doughnut Economics
Inspired by the doughnut shape, this economic model, proposed by Kate Raworth, envisions a world where humanity thrives within planetary boundaries. It calls for balancing the needs of people to live well with the capacity of the planet to support life.

The Origami Approach: Embrace the doughnut economics model in your organization. Assess how your operations impact both social foundations (such as health, education, and equity) and environmental ceilings (such as carbon emissions and resource use). By ensuring your business operates within these boundaries, you contribute to sustainable growth while safeguarding the planet.

The Blue Economy

Gunter Pauli's Blue Economy concept advocates for business practices that imitate natural systems. It encourages companies to find innovative solutions to challenges while using fewer resources.

The Origami Approach: Explore how the Blue Economy principles can be integrated into your operations. Consider how you can mimic nature's efficient processes to reduce waste, conserve energy, and create products and services that harmonize with the environment.

Employee Ownership

Studies have shown that employee-owned companies tend to focus more on long-term sustainability than their publicly traded counterparts. By giving employees a stake in the organization's success, these companies prioritize stability and prosperity over quick profits.

The Origami Approach: Investigate the feasibility of transitioning toward employee ownership or implementing employee stock ownership plans (ESOPs). Engage your workforce in the company's long-term vision, instilling a sense of ownership and commitment to sustainable growth.

The Future of Sustainable Growth

As we conclude this chapter on Sustainable Growth, I invite you to reflect upon the transformative potential of the Origami Approach. Sustainable growth is not a destination; it's a continuous journey. It's about crafting an organization that thrives in perpetuity, balancing prosperity with responsibility.

Sustainability in the Digital Age
The future of sustainable growth lies at the intersection of technology and responsibility. In a world defined by data and digital transformation, organizations have unprecedented opportunities to optimize operations, reduce waste, and enhance sustainability.

The Promise of Circular Economies
The circular economy, fueled by advances in technology, is poised to revolutionize sustainable growth. Products can be designed with recyclability in mind, and data-driven supply chains can minimize waste and optimize resource use.

The Origami Approach: Embrace the circular economy principles and leverage technology to implement sustainable practices. Consider how your organization can design products and processes that contribute to a circular economy, reducing environmental impact and promoting sustainable growth.

Ethical AI and Sustainable Decision-Making

Artificial intelligence (AI) has the potential to revolutionize decision-making processes, but ethical considerations must guide its use. Organizations must ensure that AI aligns with sustainability goals and ethical principles.

The Origami Approach: As AI becomes increasingly integrated into operations, establish ethical frameworks that prioritize sustainable decision-making. Ensure that AI-driven processes consider environmental and social impacts, contributing to the organization's long-term sustainability.

Lessons from the Past, Visions of the Future

In this exploration of sustainable growth, we've journeyed through history to uncover valuable lessons from ancient civilizations and family businesses. We've also glimpsed into the future, where technology and ethical considerations will shape the trajectory of sustainable growth.

With a clear understanding of all this, redefine your organization's purpose and role in the world. In a time when the choices we make today reverberate through generations, consider how your organization can contribute to the tapestry of sustainable growth.

Part III: The Art of Origami in Action

IIn this exhilarating phase of our journey, we delve into the application of the Origami Approach in real-world scenarios. Just as an origamist breathes life into a simple sheet of paper, we'll unfold the transformative power of these principles in the dynamic landscape of business and innovation. From tangible case studies to actionable strategies, Part III invites you to witness the magic of Business Origami taking shape in organizations worldwide. It's an opportunity to explore how these folds can reshape your own business reality and chart a course towards sustainable success.

Chapter 10: Case Studies in Business Origami

This chapter serves as a treasure trove of wisdom, inspiration, and actionable insights. In "Case Studies in Business Origami," we embark on a journey through real-life examples of organizations that have not only survived but thrived through the transformative power of the Origami Approach.

In this extensive exploration, we will dissect these case studies, peeling back the layers to reveal the essence of their success. From visionary leadership to innovative thinking, from agile strategies to adaptive cultures, each case study serves as a beacon, illuminating the path toward business excellence in an ever-evolving world.

The Power of Learning from Others

As we immerse ourselves in these stories of adaptation and reinvention, remember the ancient wisdom that "Those who cannot remember the past are condemned to repeat it." By studying the successes and challenges faced by these industry leaders, we gain invaluable insights that can inform our own business strategies and decisions.

Case Study 1: Apple Inc. - Unfolding Innovation and Customer-Centricity

Origami Insight: Just as an origamist folds paper to reveal art, Apple folds technology to craft experiences.

In the ever-evolving tech landscape, Apple Inc. stands as a testament to the power of innovation, adaptability, and customer-centricity. The company's journey from a garage startup to a global tech giant reflects the principles of the Origami Approach.

A Visionary's Aspiration

Apple's story begins with the visionary leadership of Steve Jobs. From the outset, Jobs had a clear vision—to make technology accessible, intuitive, and beautiful. He understood that technology, like paper, could be folded and molded to create something extraordinary.

Innovation Fold: Steve Jobs believed that innovation wasn't just about adding features but about simplifying complex technologies to enhance the user experience. He folded the idea of innovation into Apple's DNA.

The Unveiling of Revolutionary Products

The Macintosh: A Fold in Time

In 1984, Apple introduced the Macintosh, a computer that combined cutting-edge technology with a user-friendly interface. The Macintosh unfolded a new era of personal computing.

Customer-Centric Fold: Apple's focus on creating products that users love was evident in the Macintosh. Jobs understood that technology should adapt to people, not the other way around.

The iPod: A Harmonious Fold
The iPod, introduced in 2001, demonstrated Apple's ability to pivot and adapt to the digital music revolution. It wasn't just a music player; it was an ecosystem that folded convenience and style into one device.

Strategic Agility Fold: Apple recognized the changing landscape of the music industry and responded with agility. They adapted their strategy to cater to the emerging demand for digital music consumption.

The iPhone: A Fold in Communication
In 2007, Apple unveiled the iPhone, a device that revolutionized not just the smartphone industry but the way we communicate, work, and live. It was a fold in the fabric of daily life.

Innovative Thinking Fold: The iPhone wasn't just a phone; it was a platform for innovation. Apple fostered a culture of thinking beyond the obvious, resulting in the App Store, which unfolded endless possibilities.

The iPad: A Foldable Future

The iPad, launched in 2010, was a daring move. It folded the power of a computer into a sleek tablet, challenging the notion of what computing could be.

Adaptive Culture Fold: Apple's willingness to disrupt its own markets by introducing new products like the iPad demonstrates an adaptive culture that embraces change as a constant.

Customer-Centricity: The Fold that Holds It All

At the core of Apple's success is its unwavering commitment to customer-centricity. The company listens to its users, anticipates their needs, and folds those insights into its product design.

Origami Insight: "Apple's secret is simple—fold your focus around the customer. Understand their desires and pain points, then craft solutions that exceed their expectations."

Conclusion

Apple's journey is a masterclass in business origami. It's a story of visionary leadership, strategic agility, innovative thinking, adaptive culture, and above all, customer-centricity. Apple's ability to fold technology into art has not only shaped industries but also inspired a generation of businesses to follow the Origami Approach.

Origami Insight for Readers: "Just as Apple listens to its users, actively seek feedback and adapt your business to meet their evolving needs. Like a well-folded origami creation, your success will be a work of art."

As we unfold more case studies in the chapters to come, remember that each one offers unique insights and lessons. The Origami Approach is not a one-size-fits-all strategy but a mindset that can be tailored to fit the unique needs of any business.

Case Study 2: Alibaba Group - Navigating Global E-Commerce with Customer-Centric Origami

Origami Insight: Just as an origamist crafts intricate designs, Alibaba folds technology and innovation into global e-commerce success.

In the realm of global e-commerce, Alibaba Group has emerged as a pioneer, illustrating how an unwavering commitment to customer-centricity and innovative thinking can transform a business into a global powerhouse.

Jack Ma's Visionary Fold

Founded by Jack Ma in 1999, Alibaba began as an online marketplace connecting Chinese manufacturers to global buyers. Jack Ma envisioned a platform that would break down trade barriers and create opportunities for businesses of all sizes.

Visionary Leadership Fold: Jack Ma's vision was a fold in the traditional e-commerce landscape. He aimed to empower small and medium-sized enterprises (SMEs) and revolutionize global trade.

The Genesis of Alibaba's Success

Taobao: Fostering an Ecosystem

In 2003, Alibaba launched Taobao, a consumer-to-consumer (C2C) platform. This innovative move encouraged individuals to buy and sell online, transforming e-commerce in China.

Innovative Thinking Fold: Taobao's emergence was an innovative fold that democratized e-commerce. It encouraged innovation in e-commerce business models.

Alipay: A Secure Fold
To address payment concerns in online transactions, Alibaba introduced Alipay in 2004. This digital payment platform provided secure and convenient transactions.

Customer-Centric Fold: Alipay's introduction was a response to customer needs for secure online transactions, showing a deep understanding of user concerns.

Going Global with an Adaptive Mindset
Alibaba's global expansion was marked by the launch of AliExpress in 2010, enabling international consumers to access products from Chinese suppliers directly.

Strategic Agility Fold: The introduction of AliExpress demonstrated Alibaba's ability to adapt its platform for a global audience, responding to the growing demand for Chinese products.

Alibaba Cloud: A Digital Fold

Recognizing the importance of cloud computing, Alibaba Cloud (Aliyun) was launched in 2009. It quickly became one of the world's leading cloud providers.

Adaptive Culture Fold: Alibaba's foray into cloud services exemplifies its adaptive culture, embracing change in technology and business trends.

New Retail: A Fold in Shopping Experience

Alibaba's "New Retail" concept merges online and offline shopping experiences, reimagining retail through innovations like Hema supermarkets and Singles' Day extravaganzas.

Customer-Centric Fold: New Retail is a testament to Alibaba's customer-centric approach, delivering seamless shopping experiences that customers love.

Innovative Thinking: The Weave of Success

Central to Alibaba's journey is its culture of innovative thinking. The company continually seeks new opportunities, explores uncharted territories, and folds these ideas into reality.

Origami Insight: "Innovation isn't just about new products; it's about folding creativity into every aspect of your business, from customer experience to business models."

Conclusion

Alibaba's rise from a small apartment in Hangzhou to a global e-commerce giant reflects the Origami Approach in action. It's a story of visionary leadership, innovative thinking, customer-centricity, strategic agility, and an adaptive culture.

Origami Insight for Readers: "Like Alibaba, focus on understanding and fulfilling your customers' needs. Innovate fearlessly, adapt swiftly, and fold technology into every fold of your business."

As we continue our journey through case studies, remember that each company offers unique lessons in the Origami Approach. The principles discussed here can be folded into your own business's strategy to navigate the complex world of e-commerce and beyond.

Case Study 3: Alphabet Inc. (Google) - Unfolding Sustainable Growth and Innovation

Origami Insight: Just as an origamist shapes paper into intricate designs, Google folds innovation and sustainability into its quest for global impact.

Alphabet Inc., the parent company of Google, has redefined the tech landscape with its commitment to sustainable growth and groundbreaking innovation. Google's strategic investments in renewable energy, cloud computing, and AI technologies embody the principles of the Origami Approach.

Larry Page and Sergey Brin's Visionary Origami

Founded by Larry Page and Sergey Brin in 1998, Google's mission was to organize the world's information and make it universally accessible and useful. Their visionary leadership set the stage for Google's transformative journey.

Visionary Leadership Fold: Larry and Sergey's vision was a fold in the traditional tech landscape, focusing on democratizing information and creating products that enhance lives.

The Unveiling of Transformative Technologies

Renewable Energy: The Sustainable Fold
Google's commitment to renewable energy unfolded in 2007 with the launch of its green energy initiative. Google invested in wind and solar projects, aiming to power its operations with 100% renewable energy.

Sustainable Growth Fold: Google's investment in renewable energy was a strategic fold that aligned its growth with environmental responsibility, setting a new standard for sustainability in tech.

Google Cloud: A Skyward Fold
Google Cloud, launched in 2008, aimed to provide scalable, reliable cloud computing solutions. It sought to empower businesses with the agility to innovate and scale.

Strategic Agility Fold: Google's entry into cloud computing demonstrated its strategic agility, responding to the growing demand for cloud services and digital transformation.

AI and Machine Learning: The Intelligence Fold
Google's dedication to artificial intelligence and machine learning is evident in products like Google Assistant and TensorFlow. These innovations empower businesses and individuals with smart, data-driven solutions.

Innovative Thinking Fold: Google's pioneering work in AI showcases its commitment to innovative thinking. It continuously explores new frontiers, folding AI into diverse applications.

Sustainability: The Thread That Holds It All

At the heart of Google's success lies a profound commitment to sustainability. The company has not only invested in renewable energy but has also achieved carbon neutrality. Furthermore, it has committed to operate on 24/7 carbon-free energy by 2030.

Origami Insight: "Sustainability is not an afterthought; it should be a thread woven into the fabric of your business, just as Google has folded it into every aspect of its operations."

Conclusion

Google's journey from a search engine to a global technology conglomerate reflects the Origami Approach in practice. It's a story of visionary leadership, sustainable growth, strategic agility, innovative thinking, and an unwavering commitment to environmental responsibility.

Origami Insight for Readers: "Emulate Google's commitment to sustainability, and explore innovative technologies that can fold into your business model. The pursuit of sustainable growth is a path to success and a better future."

Case Study 4: McKinsey & Company - Navigating Consulting with Adaptive Origami

Origami Insight: Just as an origamist folds paper to create intricate designs, McKinsey folds adaptive culture and continuous learning into its consulting excellence.

McKinsey & Company, a global management consulting firm, is renowned for its emphasis on adaptive culture and relentless pursuit of learning. This case study delves into how McKinsey exemplifies the principles of the Origami Approach within the consulting industry.

A Vision of Excellence

Founded in 1926 by James O. McKinsey, the firm set out to provide innovative solutions to complex business challenges. McKinsey's vision was to foster a culture of continuous improvement and learning.

Visionary Leadership Fold: James O. McKinsey's vision laid the foundation for a consulting firm that constantly adapts and innovates to meet evolving client needs.

The Unveiling of Adaptive Culture

The McKinsey Culture: A Transformative Fold

McKinsey's culture is centered around values such as client focus, leadership, and stewardship. This culture encourages consultants to adapt, collaborate, and consistently improve.

Adaptive Culture Fold: McKinsey's emphasis on culture is a fold that encourages adaptability, ensuring that the firm can navigate a constantly changing business landscape.

Continuous Learning: The Fold of Mastery
McKinsey invests heavily in continuous learning and professional development. Consultants are encouraged to acquire new skills, embrace diverse perspectives, and stay at the forefront of industry trends.

Innovative Thinking Fold: McKinsey's commitment to learning reflects its innovative thinking. It fosters an environment where consultants continuously fold new knowledge into their practices.

Strategic Agility: A Consulting Fold
McKinsey's ability to adapt to the unique challenges faced by its clients is a testament to its strategic agility. Consultants tailor solutions to address specific client needs, ensuring maximum impact.

Strategic Agility Fold: McKinsey's adaptability is a fold that allows the firm to respond swiftly and effectively to complex business problems.

Client-Centricity: A Trusted Fold

At the heart of McKinsey's success is its unwavering commitment to client-centricity. Consultants put clients' interests first, seeking to deliver exceptional value and long-term impact.

Origami Insight: "Client-centricity is not a checkbox; it's a fold that should shape every decision and action, just as it does at McKinsey."

Conclusion

McKinsey & Company's journey from a visionary idea to a global consulting powerhouse is a compelling illustration of the Origami Approach in action. It's a story of visionary leadership, adaptive culture, continuous learning, strategic agility, and a relentless commitment to client-centricity.

Origami Insight for Readers: "Embrace a culture of adaptability and continuous learning within your organization. Just as McKinsey's consultants fold new knowledge into their work, cultivate a mindset that continually folds innovation into your business."

Case Study 5: Sony Corporation - Unfolding Innovation and Adaptability in Entertainment

Origami Insight: Just as an origamist meticulously folds paper to create art, Sony crafts innovation and adaptability into the heart of the entertainment industry.

Sony Corporation, a global leader in consumer electronics and entertainment, has long been celebrated for its innovative products and adaptability to emerging technologies. This case study explores how Sony embodies the principles of the Origami Approach within the dynamic entertainment industry.

The Vision of Founders

Founded in 1946 by Masaru Ibuka and Akio Morita, Sony's vision was to create innovative and quality products that would enrich people's lives. Their visionary leadership set the stage for Sony's transformative journey.

Visionary Leadership Fold: Masaru Ibuka and Akio Morita's vision of innovation and quality was a foundational fold that continues to shape Sony's identity.

The Unveiling of Innovative Products

The Walkman: A Musical Fold

In 1979, Sony introduced the Walkman, a portable music player that revolutionized the way people experienced music. It allowed individuals to carry their favorite tunes with them, folding music into daily life.

Innovative Thinking Fold: The Walkman was a testament to Sony's innovative thinking. It demonstrated the company's ability to fold technology into products that became cultural icons.

PlayStation: A Gaming Fold

Sony's foray into the gaming industry with the PlayStation in 1994 marked a bold move. It unfolded immersive gaming experiences, redefining interactive entertainment.

Adaptive Culture Fold: Sony's ability to enter new markets, like gaming, showcased its adaptive culture. It was willing to embrace change and pivot toward emerging opportunities.

The Convergence of Entertainment

Sony Pictures: A Cinematic Fold

Sony Pictures Entertainment, a division of Sony, has produced iconic films and television shows. It has folded storytelling and technology to create captivating entertainment experiences.

Strategic Agility Fold: Sony Pictures' ability to adapt to changing audience preferences and evolving distribution models reflects strategic agility in the entertainment industry.

Sony Music: A Harmonious Fold
Sony Music Entertainment has been instrumental in promoting global music artists. It has embraced digital platforms and streaming services, folding convenience into music consumption.

Customer-Centric Fold: Sony Music's customer-centric approach focuses on delivering music in ways that meet evolving consumer needs.

Innovation and Adaptability: The Heart of Sony

Central to Sony's success is its unwavering commitment to innovation and adaptability. The company continually seeks new opportunities, explores emerging technologies, and folds these ideas into reality.

Origami Insight: "Innovation should be at the core of your business, much like it is for Sony. Embrace new technologies and adapt to changing consumer expectations."

Conclusion

Sony Corporation's journey from its founders' vision to a global leader in entertainment showcases the Origami Approach in the entertainment industry. It's a story of visionary leadership, innovative thinking, adaptive culture, strategic agility, and a relentless commitment to customer-centricity.

Origami Insight for Readers: "Infuse innovation into your business's DNA and remain adaptable to emerging technologies. Just as Sony folds technology into entertainment, your adaptability will shape your success."

Case Study 6: Boeing - Navigating Aerospace with Strategic Agility

Origami Insight: Just as an origamist folds paper into intricate designs, Boeing crafts strategic agility into the heart of the aerospace industry.

Boeing, a leading aerospace company, has long been at the forefront of aviation innovation and adaptability. This case study delves into how Boeing exemplifies the principles of the Origami Approach within the complex aerospace industry.

The Pioneering Spirit

Founded in 1916 by William Boeing, Boeing's vision was to create innovative aviation solutions. The company's founders set the stage for Boeing's transformative journey.

Visionary Leadership Fold: William Boeing's vision of aviation innovation was a foundational fold that continues to shape Boeing's identity.

The Unveiling of Aviation Solutions

Boeing 707: A Jet-Age Fold
In 1958, Boeing introduced the 707, a groundbreaking commercial jetliner. It revolutionized air travel by making long-distance flights more efficient and accessible.

Innovative Thinking Fold: The Boeing 707 showcased the company's innovative thinking, folding jet propulsion into the future of aviation.

Boeing 747: A Global Fold

The iconic Boeing 747, introduced in 1970, transformed global air travel. It unfolded the possibilities of long-haul flights and opened new markets.

Strategic Agility Fold: Boeing's ability to innovate and create aircraft that addressed evolving market dynamics reflected its strategic agility in the aerospace industry.

The Aerospace Ecosystem

Sustainable Aviation: An Eco-Friendly Fold

Boeing has taken significant steps towards sustainable aviation. Initiatives like the Boeing ecoDemonstrator program and the development of eco-friendly aviation fuels show a commitment to reducing the environmental footprint of aviation.

Sustainable Growth Fold: Boeing's commitment to sustainable aviation is a fold that aligns growth with environmental responsibility, shaping the future of aerospace.

Space Exploration: A Cosmic Fold

Boeing's involvement in space exploration, including projects with NASA, highlights its adaptability to emerging technologies and the growing commercial space industry.

Adaptive Culture Fold: Boeing's ability to pivot toward space exploration reflects an adaptive culture open to new opportunities.

Strategic Agility: The Aerospace Navigator

At the core of Boeing's success is its strategic agility. The company continually adapts to changing market dynamics, technological advancements, and evolving customer needs.

Origami Insight: "Strategic agility should be a cornerstone of your business, as it is for Boeing. Adapt to market shifts and embrace innovative solutions."

Conclusion

Boeing's journey from its founders' vision to a global aerospace leader showcases the Origami Approach in action. It's a story of visionary leadership, innovative thinking, sustainable growth, adaptive culture, and a relentless commitment to strategic agility.

Case Study 7: Intel - Pivoting in the Tech Landscape with Agility

Origami Insight: Just as an origamist folds paper into intricate designs, Intel crafts strategic agility into its core, adapting to the ever-evolving tech landscape.

Intel, a renowned chip manufacturer, has not only thrived in the tech industry but also evolved to provide integrated solutions. This case study explores how Intel embodies the principles of the Origami Approach, demonstrating agility and adaptability in a rapidly changing tech landscape.

The Silicon Pioneer

Founded in 1968 by Robert Noyce and Gordon Moore, Intel's vision was to design and manufacture semiconductors. The company's founders set the stage for Intel's transformative journey.

Visionary Leadership Fold: Noyce and Moore's vision of semiconductor innovation was a foundational fold that continues to shape Intel's identity.

The Unveiling of Semiconductors

The Microprocessor Revolution: A Digital Fold
In 1971, Intel introduced the 4004 microprocessor, initiating the digital revolution. It unfolded the possibilities

of personal computing, changing the tech landscape forever.

Innovative Thinking Fold: The 4004 microprocessor demonstrated Intel's innovative thinking, folding computing power into compact chips.

The x86 Architecture: A Compute Fold

Intel's x86 architecture, introduced in 1978, became the industry standard for microprocessors. It unfolded the power of compatibility, enabling software to run on multiple platforms.

Strategic Agility Fold: Intel's adoption of the x86 architecture reflected strategic agility, as it responded to the evolving needs of the computing market.

The Tech Ecosystem

Integrated Solutions: A Convergence Fold

Intel has evolved beyond chip manufacturing to provide integrated solutions for computing, networking, and the cloud. Initiatives like the Intel Inside program demonstrate a commitment to a holistic tech ecosystem.

Strategic Agility Fold: Intel's pivot towards integrated solutions is a strategic fold that aligns with the changing tech landscape, ensuring relevance and innovation.

AI and Edge Computing: A Futuristic Fold

Intel has embraced emerging technologies like artificial intelligence and edge computing. Initiatives such as the Intel AI Builders program showcase its adaptability to evolving tech trends.

Innovative Thinking Fold: Intel's commitment to AI and edge computing reflects innovative thinking, folding these technologies into its offerings.

Strategic Agility: The Tech Navigator

At the core of Intel's success is its strategic agility. The company continually adapts to changing tech trends, market dynamics, and customer demands.

Origami Insight: "Strategic agility is the cornerstone of success, just as it is for Intel. Adapt to tech shifts and embrace innovation."

Conclusion

Intel's journey from a semiconductor manufacturer to a provider of integrated solutions epitomizes the Origami Approach in the tech industry. It's a story of visionary leadership, innovative thinking, strategic agility, adaptive culture, and a relentless commitment to staying relevant.

Case Study 8: Adobe - Creative Transformation through Subscription Innovation

Origami Insight: Like a skilled origamist, Adobe has folded innovation and adaptability into the heart of its operations, crafting a transformative journey

Adobe, a global leader in creative software, embarked on a bold transition to a subscription-based model. This case study explores how Adobe exemplifies the principles of the Origami Approach through creative innovation and strategic adaptation.

Creative Visionaries

Adobe was founded in 1982 by John Warnock and Charles Geschke with the vision of creating software that empowers creativity. Their visionary leadership laid the foundation for Adobe's transformative journey.

Visionary Leadership Fold: Warnock and Geschke's vision to empower creativity was a foundational fold that continues to shape Adobe's identity.

Unfolding Creative Innovation

The Creative Suite: A Digital Canvas

Adobe's Creative Suite, launched in 2003, revolutionized digital content creation. It unfolded a comprehensive toolkit for designers, photographers, and creatives worldwide.

Innovative Thinking Fold: The Creative Suite exemplified Adobe's innovative thinking, folding creative power into software solutions.

Transition to Subscription: A Paradigm Fold
Adobe's transition from traditional software licenses to a subscription-based model, Adobe Creative Cloud, marked a significant shift. It unfolded a flexible and accessible approach to creative software.

Strategic Agility Fold: Adobe's bold transition demonstrated strategic agility, adapting to changing market demands and providing ongoing value to its customers.

The Creative Ecosystem

Creative Cloud Services: A Collaborative Fold
Adobe expanded Creative Cloud to include a suite of services, fostering collaboration among creatives. It unfolded a digital ecosystem where teams can work seamlessly across devices.

Adaptive Culture Fold: Adobe's commitment to collaboration reflects an adaptive culture, embracing new ways of creative work.

AI-Powered Tools: An Artistic Fold

Adobe infused AI and machine learning into its tools, like Adobe Sensei and Adobe Firefly. It unfolded the potential to automate tasks, enhance creativity, and streamline workflows.

Innovative Thinking Fold: Adobe's integration of AI reflects innovative thinking, folding intelligent technology into creative processes.

Strategic Agility: The Creative Navigator

At the core of Adobe's success is its strategic agility. The company continually adapts to changing creative trends, technological advancements, and customer expectations.

Origami Insight: "Strategic agility is your compass to success, much like it is for Adobe. Adapt to creative shifts and embrace innovation."

Conclusion

Adobe's journey from a software company to a creative ecosystem provider showcases the Origami Approach in action. It's a story of visionary leadership, innovative thinking, strategic agility, an adaptive culture, and a relentless commitment to empowering creativity.

Chapter 11: The Challenges of Unfolding

In the world of business, much like the art of origami, transformation is a delicate yet transformative process. The Origami Approach offers a powerful framework for adaptation, but it's not without its hurdles. In this chapter, we explore the often-neglected but critical challenges that organizations face when embracing transformation.

The Pitfalls of Transformation

Pitfall 1: The Vision Vacuum

Much like starting an origami project without a clear vision, embarking on a transformation journey without visionary leadership can lead to confusion and disarray. When leaders fail to articulate a compelling vision and align the organization, transformation efforts can lose direction. According to McKinsey, 70% of transformation programs fail due to a lack of leadership alignment.

Navigational Tip: Leaders must craft an inspiring vision and ensure it permeates the organization. Effective communication and engagement are non-negotiable.

Pitfall 2: Resistance, the Unseen Enemy

Resistance to change is an age-old challenge, akin to trying to fold a stubborn piece of paper. Just as paper can resist folding, employees may resist new processes, technologies, or cultural shifts. Resistance can manifest as fear, skepticism, or inertia, acting as a significant roadblock to transformation. Prosci reports that 73% of organizations cite resistance as their biggest challenge in change initiatives.

Navigational Tip: Resistance should be anticipated and addressed proactively. Employee involvement, robust training, and clear communication of the benefits of change are key to overcoming this obstacle.

Pitfall 3: The Innovation Drought

Innovation is the lifeblood of adaptation, similar to how creativity fuels origami designs. Neglecting to foster a culture of innovation can stifle an organization's adaptability. Companies that stagnate in their innovation efforts risk becoming obsolete in today's rapidly changing business landscape.

Navigational Tip: Encourage a culture of innovation at all levels of the organization. Allocate resources for research and development, incentivize creative thinking, and create an environment where experimentation is embraced.

Pitfall 4: Myopia on Immediate Gains

Short-term thinking can blind organizations to long-term success. Focusing solely on immediate gains is akin to disregarding the broader origami design. Prioritizing quick wins at the expense of long-term sustainability can lead to failure.

Navigational Tip: Balance short-term goals with a long-term vision. Evaluate the enduring impact of decisions on the organization's future.

Pitfall 5: Neglecting Cultural Transformation
Transformation isn't just about processes and technology; it's also about culture. Just as a fold is a fundamental part of origami, cultural alignment is crucial. Neglecting cultural transformation can lead to employee disengagement and hinder the adoption of new practices.

Navigational Tip: Cultural change should be deliberate and aligned with transformation goals. Leaders should model desired behaviors, and employees should participate in shaping the new culture.

Overcoming Resistance to Change

Strategy 1: Change Leadership
Change leadership is more than just a buzzword. It involves leaders at all levels of the organization exhibiting qualities of resilience, adaptability, and empathy. An MIT Sloan Management Review study found that organizations with effective change leadership are 3.5 times more likely to outperform their peers.

Strategy 2: Empowering Change Agents
Change agents are the unsung heroes of transformation, much like skilled origamists who guide a design to life. Identifying and empowering change agents within the organization can be transformative. These individuals champion the transformation, inspire their peers, and act as conduits for feedback and concerns. A Harvard Business Review study suggests that change agents can increase the

likelihood of success in transformation efforts by as much as 60%.

Strategy 3: Open and Transparent Communication
Effective communication is the linchpin of successful change management. Transparency in conveying the reasons behind the transformation, its progress, and the expected outcomes builds trust and reduces uncertainty. A Towers Watson study found that companies with the most effective communication during change were 3.5 times more likely to outperform their peers.

Strategy 4: Data-Driven Decision-Making
Just as origami requires precision, decisions in transformation should be data-driven. Leveraging data and analytics can help organizations make informed choices. Data-driven insights provide clarity on what is working, what needs adjustment, and where resources should be allocated. A PwC survey found that 60% of respondents cited data and analytics as essential for effective change management.

Strategy 5: Cultivate Continuous Learning
Continuous learning and development initiatives are critical to equip employees with the skills and knowledge required for the transformed environment. Learning should be embedded into the organization's culture. A report by the Corporate Executive Board (CEB) found that employees in organizations with strong learning cultures are 34% more likely to feel confident navigating change.

Navigating the Unfolding Journey

Business transformation, much like the art of origami, requires a delicate touch, perseverance, and an understanding of the challenges involved. Recognizing these common pitfalls and proactively addressing them is vital for a successful transformation journey. Moreover, overcoming resistance to change demands a strategic, holistic approach that encompasses leadership, communication, data, culture, and continuous learning.

The Origami Approach provides a robust framework, but it is only as effective as the organization's ability to navigate these challenges and embrace the transformative power of adaptation.

Part IV: The Origami Masterclass

In this part, we elevate your understanding of the Origami Approach to a masterful level. You'll explore advanced strategies, intricate folds, and real-world applications that empower you to become a virtuoso of business adaptation. Prepare to embark on a transformative journey that will enable you to shape the future of your organization with precision and finesse.

Chapter 12: Precision Folding

The Art of Strategic Alignment

Strategic vs. Tactical: A Nuanced Approach

In the world of origami, the first fold sets the stage for the entire creation. Similarly, in business, strategic alignment lays the foundation for success. Yet, unlike origami, where a single piece of paper remains static, businesses operate in a dynamic landscape where strategies must continually adapt.

Strategic alignment begins by understanding the nuanced difference between strategy and tactics. Just as an origamist chooses between valley folds and mountain folds, organizations must discern when to deploy strategic maneuvers or tactical adjustments. Strategy provides the overarching direction, while tactics are the specific actions that execute the plan.

Successful companies, such as Apple, illustrate the significance of this nuance. Apple's strategic focus on innovation, premium design, and user experience underpins its tactical decisions in product development, marketing, and retail.

The Role of Clear Objectives

Just as an origami pattern relies on precise instructions, business alignment requires clear objectives. Setting objectives provides a roadmap for the organization, guiding decision-making and resource allocation. Objectives serve as the guiding stars, helping companies navigate a sea of choices.

Amazon, for example, defines its objectives with remarkable clarity. Jeff Bezos's famous "customer obsession" mantra permeates the company's culture, influencing strategic and tactical choices. This relentless focus on customer satisfaction shapes their innovative strategies.

Aligning Business Units and Departments

In origami, each fold contributes to the overall design's harmony. Similarly, within an organization, aligning various business units and departments ensures a coordinated effort. Misalignment can lead to inefficiencies, duplicated efforts, and conflicts.

A classic example of successful alignment is seen in Toyota's lean production system. Every aspect of Toyota's operations, from manufacturing to supply chain management, is harmonized through principles like just-in-time production and continuous improvement.

Measuring Alignment Effectiveness

In the intricate world of origami, measuring precision can be a matter of millimeters. Similarly, in business, measuring alignment effectiveness is essential for fine-tuning strategies. Key Performance Indicators (KPIs) serve as the metrics to gauge alignment success.

Take the example of Google's parent company, Alphabet Inc. Through subsidiary companies like Waymo for autonomous vehicles and Verily for life sciences, Alphabet's alignment with its core mission to organize the world's information is assessed through KPIs related to innovation, market leadership, and profitability.

Mastering Organizational Design

In origami, different folds and angles shape the final design's aesthetics. Similarly, in business, organizational design influences the company's structure, culture, and efficiency. The modern business landscape requires a departure from traditional hierarchies.

Beyond the Traditional Hierarchy

Traditional organizational hierarchies resemble a simple origami design. While they may serve a purpose, they lack the complexity needed to navigate today's intricate business world. Forward-thinking companies explore alternative organizational structures.

One such innovator is Spotify, which adopted a "tribe" model. This decentralized structure encourages autonomy and creativity among teams, fostering innovation in a rapidly evolving music streaming industry.

Agile Organizational Structures

Just as origami can take on various forms, agile organizational structures adapt to changing conditions. Agility involves a fundamental shift in how companies operate. Agile organizations respond rapidly to market shifts, customer demands, and emerging opportunities.

Netflix is a prime example of agility in action. From DVD rentals to streaming, and from content distribution to content creation, Netflix continually transforms itself. Its agile structure allows for quick adjustments and bold strategic shifts.

The Impact of Digital Transformation

In origami, the introduction of digital tools has expanded design possibilities. Likewise, digital transformation revolutionizes how businesses operate. It impacts everything from customer interactions to supply chain management.

Amazon's pioneering use of digital technologies, like AI-powered recommendations and robotic fulfillment centers, exemplifies the transformative impact of digitalization. These innovations amplify customer experience and operational efficiency.

Design Thinking in Organizational Design

Origami artists often incorporate design thinking, a creative problem-solving approach, into their projects. Similarly, organizations employ design thinking to reimagine their structures and processes. Design thinking prioritizes empathy, ideation, and iterative prototyping.

IDEO, a global design consultancy, exemplifies design thinking. Their organizational design prioritizes collaboration, experimentation, and a user-centric approach. It's a philosophy deeply embedded in their culture, driving innovation across industries.

Optimizing Resource Allocation

Resource Scarcity and Creative Solutions

Much like origami artists must make the most of limited paper, businesses often face the challenge of resource scarcity. In a competitive landscape, the efficient allocation of resources becomes a strategic necessity.
Resource scarcity isn't always about financial constraints; it can include limitations on time, talent, or technology. Innovative companies don't view scarcity as a roadblock but as an opportunity to exercise creative solutions.

One remarkable example is Airbnb. In its early stages, Airbnb faced resource constraints compared to established hotel chains. However, the company embraced the sharing economy concept, turning ordinary people into hosts. This resource-efficient approach not only saved the company money but also created a unique selling point that disrupted the traditional hospitality industry.

Agile Budgeting Techniques

Traditional budgeting often involves static, year-long plans that leave little room for flexibility. In the dynamic business landscape, this approach can be restrictive. Agile budgeting techniques offer a solution.

Agile budgeting allows companies to adapt their financial plans in response to changing conditions. It involves shorter budgeting cycles, continuous monitoring, and the ability to reallocate resources swiftly. This approach aligns financial planning with the rapid pace of change in today's markets.

A standout example is Spotify's adoption of rolling forecasts. Instead of committing to a fixed annual budget, Spotify adjusts its financial plans every quarter based on evolving market conditions and strategic priorities. This agile budgeting practice allows them to remain responsive to shifts in the music streaming industry.

Data-Driven Resource Allocation
Origami is a precise art that relies on measurements and calculations. Similarly, in business, data-driven resource allocation hinges on gathering and analyzing relevant information to inform decisions.

Companies like Netflix exemplify this approach. They use advanced analytics to determine where to invest in content creation, which helps them tailor their offerings to their global audience's preferences. By analyzing user data, they allocate resources to produce content that resonates with viewers, leading to a more engaged audience and increased retention rates.

Balancing Short-Term and Long-Term Investments
In origami, balance is crucial to achieving the desired outcome. Similarly, in business, achieving equilibrium

between short-term and long-term investments is essential for sustainable growth.

Amazon provides an insightful example of this balance. While they continually invest in long-term ventures like developing their own delivery network and building data centers to support AWS, they also focus on delivering short-term results through their e-commerce business. This dual focus allows Amazon to sustain growth over time while remaining adaptable to changing market dynamics.

Leveraging Technology Synergy

Integrating Disparate Systems
In origami, different folds come together to create a unified design. Similarly, businesses often rely on a variety of systems, both old and new. Integrating these disparate systems can unlock efficiencies and insights.

Salesforce, a leading CRM provider, exemplifies the integration of disparate systems. Their platform seamlessly connects sales, marketing, and customer service data. This integration empowers businesses to gain a 360-degree view of their customers, improving decision-making and enhancing customer experiences.

The Role of Artificial Intelligence
Just as origami artists use precise techniques, artificial intelligence (AI) employs algorithms and data to make precise predictions and automate processes. In business, AI is transforming industries by optimizing operations,

personalizing customer experiences, and driving innovation.

Netflix leverages AI to personalize content recommendations, keeping viewers engaged and reducing churn. By analyzing user behavior and preferences, Netflix's AI algorithms suggest content tailored to individual tastes, ultimately enhancing the customer experience.

Data Interoperability and Insights

Origami requires a deep understanding of how different folds interact. In the business context, achieving data interoperability—seamless data exchange between systems—is crucial for harnessing valuable insights.

Healthcare organizations like Mayo Clinic recognize the importance of data interoperability. They've implemented Health IT systems that allow for the secure sharing of patient data among providers. This interoperability enhances patient care, streamlines processes, and leads to better health outcomes.

Maximizing ROI on Technology Investments

Origami artists maximize their use of paper to create intricate designs. Similarly, businesses aim to maximize their return on investment (ROI) on technology expenditures. This involves careful planning, strategic implementation, and ongoing evaluation.

Tesla, known for its electric vehicles and clean energy solutions, exemplifies this principle. By strategically investing in electric vehicle technology and solar energy solutions, Tesla has not only disrupted traditional industries but also demonstrated how technology investments can yield long-term ROI while contributing to environmental sustainability.

We've now explored the art of efficient resource allocation, agile budgeting techniques, data-driven resource allocation, and the balance between short-term and long-term investments. We've also delved into the fusion of tech ecosystems, the role of artificial intelligence, data interoperability, and maximizing ROI on technology investments. These folds of knowledge provide the tools and insights you need to master the Origami Approach and sculpt your business's success.

Chapter 13: Adaptive Innovation

Cultivating a Culture of Innovation

Building an Innovation Ecosystem

In origami, the artist's ecosystem consists of creativity, paper, and technique. In business, fostering innovation requires creating an environment where creativity, resources, and methodology converge.

Companies like Google epitomize the development of innovation ecosystems. Google's "20% time" policy allows employees to dedicate a portion of their work hours to projects of their choice. This policy fosters creativity, leading to innovations such as Gmail and Google News.

Encouraging Intrapreneurship

Origami artists often experiment with unconventional folds to create unique designs. Similarly, intrapreneurs within organizations explore unconventional ideas to drive innovation. Encouraging intrapreneurship involves empowering employees to act as entrepreneurs within the company.

3M, the innovation powerhouse, famously allows employees to spend 15% of their work time on projects outside their regular responsibilities. This freedom has led to breakthroughs like Post-it Notes and Scotchgard.

Innovation Metrics and KPIs

Just as origami progress can be measured by precision and complexity, innovation success in business relies on metrics and key performance indicators (KPIs). These measurements ensure that innovation aligns with strategic goals.

Apple, known for its innovative products, uses KPIs like customer satisfaction, market share, and revenue growth to evaluate the impact of its innovations. These metrics help Apple gauge whether its innovations resonate with consumers and drive business results.

Open Innovation and Collaboration

Collaborative Ecosystems

Origami can be a solitary art, but its beauty often emerges from collaboration. Similarly, open innovation thrives on collaboration with external partners, customers, and even competitors.

General Electric (GE) embraces open innovation through initiatives like GE Ventures and GE Garages. By collaborating with startups and external innovators, GE leverages external expertise and accelerates the development of new technologies.

Crowdsourcing Innovation

Origami enthusiasts worldwide share their creations and techniques through online communities. Crowdsourcing innovation harnesses the collective intelligence of a diverse crowd to solve complex problems.

LEGO, the beloved toy company, crowdsources ideas from its community through LEGO Ideas. Fans submit designs, and if they garner enough support, LEGO turns them into commercial sets. This approach ensures that LEGO's product offerings align with customer preferences.

Industry-Academia Partnerships
Collaboration between industry and academia can be a wellspring of innovation. Just as origami artists learn from masters, businesses benefit from partnerships with educational institutions.

MIT's Industrial Liaison Program (ILP) serves as a prime example. It facilitates collaborations between MIT researchers and industry leaders, fostering breakthrough innovations in various sectors.

Managing Intellectual Property
In origami, knowledge is passed down through generations. In business, managing intellectual property (IP) is crucial to protect innovations and drive value.

IBM, a pioneer in IP management, holds one of the world's largest patent portfolios. Their strategic approach to IP not only safeguards their innovations but also generates revenue through licensing agreements.

The Concept of "Perpetual Beta"

Customer-Centric Product Development

Origami artists refine their designs based on aesthetic preferences. Similarly, businesses continuously refine products based on customer feedback.

Adobe embraces perpetual beta through its Creative Cloud platform. Features and updates are released incrementally, allowing Adobe to gather user feedback and tailor their products to customer needs.

Rapid Prototyping and Iteration

Origami artists prototype their designs before creating the final piece. In business, rapid prototyping and iteration accelerate product development.

IDEO, a design and innovation consultancy, employs rapid prototyping to bring ideas to life quickly. This iterative approach allows them to test and refine concepts efficiently.

User Feedback as the Compass

Origami artists rely on feedback to enhance their craft. In business, user feedback serves as a compass, guiding product development.

Amazon's customer reviews and ratings system is a prime example. User-generated feedback influences product rankings and informs Amazon's product development decisions.

Disruptive Innovation Strategies

In origami, disruption occurs when traditional folds are challenged. In business, disruptive innovation strategies challenge established norms and industries.

Netflix disrupted the traditional television and film industry by offering on-demand streaming. The company's innovative subscription-based model and original content production have reshaped the entertainment landscape.

Identifying Disruption Opportunities

Innovator's Dilemma: A Guide

Disruption is often the result of new technologies or business models challenging established industries. Clayton Christensen's Innovator's Dilemma framework offers valuable insights for identifying, understanding, and navigating disruptive opportunities.

Understanding the Innovator's Dilemma: The Innovator's Dilemma describes the challenge faced by established companies when confronted with disruptive innovations. These innovations typically target underserved or overlooked segments of the market. Established companies, focused on serving their existing customer base, may dismiss these innovations as niche or unprofitable. However, disruptive innovations have the potential to redefine industries. They often start with lower performance and cater to customers who prioritize different attributes. Over time, they improve and gradually capture larger market segments. Just as an origami artist starts with a flat sheet of paper and transforms it into intricate folds, identifying disruption opportunities involves recognizing the potential for transformation within industries and markets. Embracing disruption requires a mindset shift akin to unfolding the paper's potential.

Embracing Disruption as an Opportunity

To identify disruption opportunities, businesses should adopt a proactive approach. This involves:

Market Monitoring

Continuously monitor the broader market for emerging technologies and business models. Recognize that disruptive opportunities may initially appear as niche markets.

Customer-Centric Research

Understand the evolving needs and preferences of your target customers. Pay attention to segments that may be underserved or willing to adopt new solutions.

Experimentation

Encourage a culture of experimentation within your organization. Allocate resources to explore and test innovative ideas, even if they initially seem peripheral to your core business.

Strategic Partnerships

Collaborate with startups, research institutions, and industry disruptors. These partnerships can provide insights into emerging trends and technologies.

5. Risk Tolerance: Recognize that disruptive ventures often involve higher risks. Be willing to embrace calculated risks and allocate resources to explore new opportunities.

Navigating Regulatory Challenges

Regulatory challenges can be as intricate as an origami masterpiece. Like an origamist who meticulously follows folding instructions, businesses must navigate regulatory frameworks with precision. Disruptive innovations frequently encounter regulatory obstacles, as existing regulations may not adequately address new technologies or business models. Navigating these challenges is a crucial aspect of seizing disruption opportunities.

Engaging with Regulators
Establish open lines of communication with regulatory authorities. Engage in proactive dialogue to educate regulators about the potential benefits of the disruptive innovation and to address any concerns they may have.

Legal and Compliance Expertise
Develop in-house expertise or seek external legal counsel well-versed in the relevant regulatory landscape. Understanding and complying with regulations is essential to mitigate legal risks.

Advocacy and Industry Collaboration
Collaborate with industry associations and peers to advocate for regulatory changes that accommodate the

disruptive innovation. Collective efforts can have a more significant impact on policy reforms.

Scaling Disruptive Ventures
Once a disruption opportunity is identified and regulatory challenges are navigated, the next challenge is scaling the disruptive venture effectively.

Resource Allocation
Allocate resources strategically to support the scaling of the disruptive venture. Consider the following:

- **Talent Acquisition:** Attract and retain talent with expertise in the disruptive technology or business model.
- **Investment:** Secure funding and investment to fuel the growth of the disruptive venture.
- **Infrastructure:** Build the necessary infrastructure and capabilities to support scalability.

Market Expansion
Gradually expand the disruptive venture's market presence. Initially, focus on the niche or underserved segment and then broaden the reach as the offering matures.

Continuous Innovation
Continue to innovate and refine the disruptive venture. Feedback from early adopters and customers should inform ongoing improvements.

Risk Management

Acknowledge that scaling disruptive ventures involves inherent risks. Implement risk management strategies to address potential challenges and uncertainties.

By embracing the origami perspective on disruption, businesses can gain valuable insights into the transformative journey of identifying opportunities, navigating regulatory challenges, and scaling ventures. Much like origami's artistry lies in the folds, the art of business lies in the strategic and precise unfolding of disruptive potential.

Chapter 14: Strategic Agility: Crafting a Strategy as Agile as Origami

In the intricate world of origami, the ability to adapt to new folds and challenges is paramount. Similarly, in the realm of business, strategic agility is the key to not just surviving but thriving in an ever-changing landscape. Just as an origami artist deftly adjusts their folds to create a new masterpiece, organizations must continually adapt their strategies to seize emerging opportunities.

Dynamic Strategy Formulation

Agile Strategy Development
In a business landscape where change is constant, organizations need to approach strategy development with the agility of an origami artist crafting a unique design. Agile strategy development involves embracing flexibility and responsiveness.

Much like the precise folds in origami, agile strategy development requires organizations to break down complex strategies into manageable components. This allows for quicker adjustments as circumstances evolve. Agile organizations empower teams to adapt and make strategic decisions in real-time, mirroring the creativity and adaptability seen in origami.

Continuous Strategy Evaluation
The path to strategic agility is akin to the evolution of an origami design, with constant evaluation and adjustment.

Organizations must continuously assess the effectiveness of their strategies and make necessary modifications. Continuous strategy evaluation involves the regular review of key performance indicators (KPIs) and market dynamics. Organizations that excel in this area are like origami artists who, with each fold, refine their creation to achieve perfection. They leverage data and feedback to refine their strategies, ensuring they remain aligned with their goals.

Scenario-Based Planning
Just as origami artists envision various outcomes during the folding process, organizations must engage in scenario-based planning. This involves preparing for different potential futures and having strategies in place for each. Scenario-based planning allows organizations to respond rapidly to changing circumstances, much like an origami artist adjusting their folds based on the paper's characteristics. By anticipating a range of scenarios, businesses can adapt swiftly to unexpected challenges or opportunities.

Strategy as a Living Document
Origami creations are never truly static; they can be unfolded and reshaped. Similarly, modern strategies should not be rigid documents but living entities that evolve over time.

Treating strategy as a living document means being open to constant refinement and adaptation. Organizations must be willing to pivot when needed, just as an origami artist might change their design mid-fold to achieve a better result. This dynamic approach to strategy ensures relevance and effectiveness in a rapidly changing world.

Strategic Partnerships and Alliances

The Power of Ecosystems

Strategic agility extends beyond individual organizations; it's also about the ability to form dynamic partnerships and alliances, much like origami pieces coming together to create a larger, more intricate design.

In today's interconnected world, the power of ecosystems cannot be overstated. Ecosystems bring together a diverse array of partners, each contributing unique strengths to achieve common goals. Just as origami pieces fit together to form a cohesive whole, ecosystems leverage complementary capabilities to drive innovation and growth.

Building Strategic Partnerships

Forming strategic partnerships is akin to selecting the right pieces for an origami creation. Organizations must carefully choose partners whose strengths align with their objectives.

Strategic partnerships are not merely transactional; they are collaborations deeply rooted in shared values and mutual benefit. Like origami artists who collaborate with others to create stunning works of art, businesses engage in partnerships to enhance their capabilities and access new markets.

Managing Alliance Risks
Just as an origami creation can be delicate, strategic alliances can be fragile without proper care. Managing alliance risks involves nurturing these partnerships to ensure longevity and success.

Organizations that excel in this area are like origami artists who meticulously preserve their delicate creations. They invest in relationship-building, communication, and conflict resolution. By proactively addressing challenges, they prevent alliances from unraveling.

Collaborative Growth Strategies
Origami artists often collaborate with others to create intricate designs. Similarly, businesses engage in collaborative growth strategies to achieve shared objectives.

Collaborative growth strategies involve joint ventures, co-development, or co-marketing efforts. These strategies leverage the strengths of multiple parties to achieve results that may not be attainable individually, much like origami artists teaming up to create larger, more intricate pieces.

Global Expansion and Market Entry

Entering Emerging Markets
Expanding into emerging markets is like venturing into uncharted origami territory. It's a journey filled with unique challenges and opportunities that require careful consideration.

Organizations must adapt their strategies to navigate cultural differences, regulatory complexities, and evolving consumer behaviors in these markets. Just as origami artists study and respect different paper textures and qualities, businesses must understand the intricacies of each market they enter.

Navigating Cultural Differences
Understanding and respecting cultural differences is essential for global success, much like an origami artist respecting the characteristics of different types of paper. When expanding internationally, organizations must adapt their strategies to align with the cultural norms and values of each market. This includes tailoring products, marketing, and communication approaches to resonate with local audiences.

Global Supply Chain Agility
In a world where supply chains span continents, supply chain agility is essential, much like the precision needed in origami folds.

Agile supply chains allow organizations to respond swiftly to disruptions, whether caused by natural disasters or geopolitical events. They ensure that products reach customers efficiently, regardless of the global location. This agility mirrors the adaptability of origami, where each fold is carefully executed to achieve the desired outcome.

Regulatory Compliance Across Borders

Just as origami artists follow established guidelines for creating specific designs, organizations must adhere to regulatory requirements when operating in multiple countries.

Navigating regulatory compliance across borders demands meticulous planning and adherence to local laws and regulations. Businesses must adapt their strategies to ensure compliance while maintaining operational efficiency.

Portfolio Diversification and Innovation
Building an Innovation Portfolio

Creating an innovation portfolio is akin to an origami artist experimenting with different designs and techniques. It involves strategically diversifying investments in innovation.

Innovation portfolios encompass a range of projects, from incremental improvements to disruptive innovations. Like origami artists exploring various styles, organizations allocate resources to initiatives that offer different levels of risk and potential reward.

Balancing Risk and Innovation

Just as origami artists assess the complexity of a design before embarking on it, organizations must evaluate the risks associated with their innovation portfolio.

Balancing risk and innovation involves making calculated bets on emerging technologies, markets, and business models. Organizations aim to strike a balance between pursuing transformative innovations and safeguarding their core operations.

Strategic Investments in Disruptive Ventures

Investing in disruptive ventures is like an origami artist pushing the boundaries of traditional designs. Organizations allocate resources to initiatives that have the potential to reshape industries.

These strategic investments often involve exploring emerging technologies or entering new markets. Like origami artists who embrace unconventional approaches, businesses pursue disruptive ventures to position themselves as industry leaders.

Measuring Portfolio Performance

Just as origami artists assess the quality of their creations, organizations must measure the performance of their innovation portfolio.

Key performance indicators (KPIs) and metrics help organizations gauge the impact of their innovation efforts. Regular assessments allow for adjustments, ensuring that resources are allocated to initiatives with the highest potential for success.

In the world of strategic agility, organizations must master the art of adaptability, much like origami artists who create intricate designs through precision and flexibility. By embracing agile strategy development, forming strategic partnerships, expanding globally with cultural sensitivity, and managing innovation portfolios, businesses can achieve strategic agility that rivals the beauty and complexity of origami.

Chapter 15: Transformational Technology: Orchestrating the Future

In the ever-evolving landscape of business, technology is the brush that paints the future. Just as origami artists harness the power of their tools to craft intricate designs, organizations must leverage transformative technologies to stay ahead. In this chapter, we delve into the world of Transformational Technology, exploring how Industry 4.0 and beyond, smart manufacturing, IoT, blockchain, augmented and virtual reality, artificial intelligence (AI), and machine learning are reshaping industries.

Next-Generation Digital Transformation

Industry 4.0 and Beyond

The concept of Industry 4.0 is akin to origami's precision, where every fold matters. It represents the integration of digital technologies into all aspects of business operations. From advanced robotics to the Internet of Things (IoT) and big data analytics, Industry 4.0 is revolutionizing manufacturing.

By connecting machines and systems, businesses gain real-time insights into their operations, much like origami artists meticulously adjust each fold. These insights enable predictive maintenance, optimized production processes, and cost savings. Industry 4.0 is transforming factories into intelligent, self-optimizing ecosystems.

Smart Manufacturing and IoT

Smart manufacturing, an integral part of Industry 4.0, is akin to origami's elegance in simplicity. It leverages IoT devices and sensors to collect and analyze data from the manufacturing floor. This data fuels decision-making, process optimization, and product quality improvements. Much like origami artists who require precise tools, smart manufacturing relies on interconnected sensors, data analytics, and automation. The result is enhanced productivity, reduced downtime, and the ability to respond swiftly to changing market demands.

Blockchain for Supply Chain Transparency
Blockchain, with its transparency and immutability, is akin to origami's precision in design. It's transforming supply chains by providing end-to-end visibility and traceability. Through blockchain, organizations can track products from their source to the end consumer. This technology ensures the authenticity of goods and mitigates issues like counterfeiting. Much like origami artists who meticulously document their folding process, businesses record every transaction on the blockchain, creating an unalterable ledger.

Augmented and Virtual Reality Applications
Augmented reality (AR) and virtual reality (VR) applications are like origami's ability to create immersive experiences. They're redefining industries from gaming to education and healthcare.

In the business world, AR enhances training and maintenance processes. VR facilitates virtual tours and product simulations. These technologies create engaging customer experiences, much like origami creations that captivate observers.

Artificial Intelligence and Machine Learning
Artificial intelligence and machine learning are the brains behind modern technology, mirroring origami artists' intricate thought processes.
AI algorithms analyze vast datasets, uncovering patterns and insights. Machine learning models predict customer behavior, streamline operations, and enable personalized marketing. Organizations that harness the power of AI and machine learning fold insights into their strategies, gaining a competitive edge.

AI for Predictive Analytics

Machine Learning in Product Development
Machine learning in product development is like origami's creative exploration of form and function. It empowers businesses to design products that meet evolving customer needs.

By analyzing customer feedback and market trends, machine learning guides product design. Organizations fold customer preferences and real-time data into their development process, ensuring products remain relevant.

AI in Customer Experience

AI's role in enhancing customer experiences is akin to origami's ability to evoke emotions through its design. It's reshaping customer interactions across industries.

Chatbots provide instant support, personalization engines tailor recommendations, and sentiment analysis gauges customer satisfaction. AI transforms customer experiences into seamless, personalized journeys, akin to origami creations that evoke wonder and delight.

Ethical AI and Bias Mitigation

Ethical AI and bias mitigation reflect origami's careful attention to detail. They're critical components of responsible technology use.

To gain trust, organizations must ensure their AI systems are free from biases that could lead to unfair decisions. Much like origami artists who strive for perfection in every fold, businesses refine their algorithms to align with ethical principles.

Quantum Computing and Advanced Data Analytics

Quantum computing's potential is like origami's endless possibilities, offering computational power beyond classical computing.

Quantum computers process vast datasets and solve complex problems with unprecedented speed. They enable advanced data analytics that uncover insights hidden in mountains of information. Organizations that harness quantum computing embrace origami's limitless creativity in problem-solving.

Quantum Computing's Potential

Leveraging Big Data and Advanced Analytics
Quantum computing's power to process vast datasets is akin to origami's precision in creating intricate designs. It's a game-changer for businesses seeking insights from massive data volumes.

By harnessing quantum computing, organizations can perform complex simulations, optimize supply chains, and unlock new scientific discoveries. It's a bit like origami artists who explore intricate designs by folding and unfolding their creations.

Quantum-Safe Cryptography
Quantum computing's potential to break existing cryptographic systems is a challenge that requires origami-like adaptability.

To protect sensitive data from future quantum threats, organizations are exploring quantum-safe cryptography. This is similar to origami artists who adapt their techniques to create new designs. Quantum-safe cryptography ensures data security in an era when quantum computers could crack traditional encryption.

Industries Transformed by Quantum

Quantum computing's impact on industries is like origami's influence on art—profound and far-reaching. Industries like pharmaceuticals, finance, and logistics benefit from quantum computing's optimization capabilities. Complex calculations for drug discovery, financial risk assessment, and route optimization become faster and more accurate. It's as if origami artists could transform not only paper but entire art forms.

Cybersecurity and Resilience
In the digital age, cybersecurity and resilience are like origami's ability to withstand time and change. They protect businesses from threats and ensure continuity. The Zero Trust Architecture is a cybersecurity approach that assumes no one is implicitly trusted, much like origami artists who carefully inspect each fold. It enforces strict access controls, reducing the attack surface. Cybersecurity becomes an enabler of business, just as origami's resilience enables complex creations.

The cybersecurity landscape mirrors origami's precision. Protecting digital assets requires constant vigilance, proactive measures, and adaptive strategies. Organizations that fold cybersecurity into their DNA ensure the longevity and security of their digital operations.

In the world of transformational technology, businesses must embrace change with the precision and adaptability of origami artists. Whether adopting Industry 4.0, harnessing the potential of quantum computing, or fortifying cybersecurity, organizations that master these technologies will craft a future that unfolds with innovation and resilience.

Chapter 16: Sustainable Growth Strategies

In a world where the art of business is ever-evolving, sustainability stands as the cornerstone of lasting success. This chapter explores Sustainable Growth Strategies, where businesses transcend mere profitability to foster a holistic approach, much like the intricate folds of origami that create a harmonious whole.

Sustainability as a Competitive Advantage

The Business Case for Sustainability
Sustainability isn't just a trend; it's the bedrock of competitive advantage. Organizations that embrace sustainability enjoy a range of benefits, from cost savings through resource efficiency to enhanced brand reputation. The analogy to origami lies in the careful, deliberate planning that goes into each fold—a plan that leads to a better outcome.

Let's consider Tesla, an exemplar of sustainability in the automotive industry. Their electric vehicles represent a commitment to reducing carbon emissions. While competitors initially hesitated to fold sustainability into their strategies, Tesla's success demonstrated the business case for going green. As a result, traditional automakers now pivot towards electric vehicles to stay competitive.

Circular Economy Practices

The circular economy, reminiscent of the intricate folds of origami, is about maximizing the utility of resources. It challenges the linear "take, make, dispose" model by promoting resource efficiency and waste reduction.

Companies like Patagonia, the outdoor clothing brand, exemplify circular economy practices. They encourage customers to repair and recycle their clothing, aligning with sustainability principles. Patagonia's approach mirrors the essence of origami, where a single sheet of paper is transformed into something greater.

Environmental, Social, and Governance (ESG) Metrics

Much like origami's precision, ESG metrics provide a structured framework for evaluating a company's sustainability efforts. ESG factors encompass environmental, social, and governance aspects, enabling investors and stakeholders to assess a company's impact beyond financial performance. Companies like Unilever have woven ESG metrics into their sustainability fabric. They've set ambitious goals to reduce their environmental footprint and enhance social impact.

Sustainability Reporting and Transparency

Transparency is the hallmark of both origami and sustainability. Reporting on sustainability efforts, akin to documenting each fold, builds trust and accountability.

IKEA, once again, shines in this aspect. They publish an annual sustainability report, detailing their progress in various sustainability initiatives. This transparency not only builds trust but also aligns with the ethos of origami, where each fold is a transparent step towards a beautiful outcome.

Inclusive and Responsible Business Practices
Inclusivity and responsibility are the heart of sustainable growth, much like origami's ability to create unity from diversity. Businesses that embrace inclusivity foster a culture of innovation and social responsibility.

Salesforce, a leader in customer relationship management, champions inclusivity through their diverse workforce and philanthropic efforts. They've pledged 1% of their equity, product, and employee time to improve communities worldwide. Salesforce's approach echoes origami's transformative power when disparate elements come together cohesively.

Diversity and Inclusion in the Workplace

Ethical Supply Chain Management
Ethical supply chains are like origami's seamless folds— they create a unified and ethical whole. Businesses are increasingly scrutinized for their supply chain practices, demanding ethical treatment of workers and responsible sourcing.

Nike, a global sportswear giant, learned this the hard way when they faced allegations of sweatshop labor. They subsequently transformed their supply chain, implementing rigorous labor standards and monitoring mechanisms.

Nike's journey mirrors origami's ability to turn a disordered sheet into a structured masterpiece.

Social Responsibility Initiatives
Businesses are increasingly folding social responsibility into their strategies, going beyond profit to create a positive impact. Like origami artists who shape paper into art, companies craft social initiatives that make a difference.

Unilever's "Sustainable Living Plan" demonstrates this commitment. They aim to improve the health and well-being of over one billion people while reducing their environmental footprint. Unilever's approach mirrors origami's transformative power when thoughtful folds create something beautiful.

Community Engagement and Impact
Community engagement is akin to origami's connection with observers. Businesses that engage with and positively impact their communities foster goodwill and brand loyalty.

Microsoft's Airband Initiative is a prime example. They use technology to extend internet access to underserved communities, bridging the digital divide. Microsoft's engagement unfolds like origami's storytelling, creating a narrative of positive change.

Green Technologies and Eco-Innovation
Much like origami's creative use of paper, businesses embrace green technologies and eco-innovation to shape a sustainable future.

Tesla's innovations in electric vehicles (EVs) and energy storage demonstrate eco-innovation. Their EVs have redefined the automobile industry, and their energy products promote renewable energy adoption. Tesla's approach echoes origami's creativity when paper is transformed into art.

Sustainable Transportation Solutions
In the pursuit of sustainable growth, transportation plays a pivotal role. Electric vehicles (EVs) and alternative fuels represent folds in the sustainable transportation landscape.

Electric car manufacturer, Rivian, is an emblem of innovation in sustainable transportation. Their EVs combine adventure and sustainability, unfolding a new narrative for eco-conscious travelers. Much like origami artists, Rivian meticulously crafts each detail of their vehicles.

Eco-Friendly Materials and Packaging

Eco-friendly materials and packaging represent the conscientious folds of sustainability. Businesses explore alternatives that minimize environmental impact.

Uncommon Goods, an online retailer, showcases eco-friendly products and packaging. Their commitment to sustainable sourcing and recyclable packaging aligns with eco-innovation. It's as if each product comes wrapped in a sustainable origami embrace.

Sustainable Agriculture and Food Systems
Sustainability extends to our plates, with sustainable agriculture and food systems folding into our diets.

Beyond Meat, a plant-based meat producer, challenges traditional agriculture with sustainable alternatives. Their products cater to environmentally conscious consumers, reflecting a fold towards sustainable eating habits. Much like origami's metamorphosis of paper, Beyond Meat transforms the culinary landscape.

Responsible Growth and Stakeholder Value
Responsible growth is the essence of sustainability, much like origami's delicate balance of aesthetics and structure. Businesses must grow while creating value for stakeholders and minimizing negative impacts.

In the realm of sustainable growth, businesses must fold together profitability and responsibility. Much like origami artists who craft intricate designs, these businesses create a sustainable masterpiece—one that balances growth with responsibility, inclusivity, and innovation.

This chapter unfolds the multifaceted nature of sustainable growth strategies. As you delve into these folds of sustainability, remember that, much like origami, the beauty lies in the meticulous craftsmanship—the deliberate and thoughtful approach to creating something extraordinary.

The Endless Possibilities of Business Origami

In the journey we've undertaken, we've unfolded the delicate art of Business Origami, where strategic folds and creative creases shape the landscape of business success. Now, as we reach the culmination of our exploration, we find ourselves at a crossroads—a place where the possibilities are endless, and the artistry of origami meets the pragmatism of business strategy.

The Power of the Unfolded

Much like a skilled origamist, you have learned to navigate the intricacies of visionary leadership, strategic agility, innovative thinking, adaptive culture, customer-centricity, and resilient operations. These folds are not mere concepts but tools in your arsenal, capable of transforming your organization.

Consider how visionary leadership can inspire your teams to reach new heights, much like an origami artist envisions the final form while working with a flat sheet of paper. Your leadership, too, shapes the future of your business, envisioning a thriving, adaptive, and innovative organization.

The Intersection of Theory and Reality

In this book, you've encountered countless examples of companies that have successfully applied origami-inspired strategies. However, the real beauty of Business Origami is its adaptability to any industry or organization, and the endless opportunities for customization.

Imagine the possibilities of Business Origami in your context:

Small and Mid-sized Enterprises (SMEs)
The agility and adaptability championed by Business Origami can level the playing field for SMEs, allowing them to compete with larger corporations.

Nonprofits and NGOs
In the world of social impact, the principles of adaptive culture and resilience can mean the difference between success and stagnation.

Government Agencies
Even the public sector can benefit from the innovative thinking and strategic agility of Business Origami to better serve citizens.

Startups
Visionary leadership and customer-centricity are pivotal for startups to thrive in an ever-changing landscape.

The point is that Business Origami is not a one-size-fits-all approach. It's a philosophy, a mindset, and a toolkit for anyone seeking to navigate the complexities of the business world.

Crafting Your Origami Strategy

So, how can you apply the principles of Business Origami to your organization's unique challenges and opportunities? Let's explore some essential steps:

Embrace the Origami Mindset

Much like a seasoned origamist, embrace a mindset of precision, patience, and creativity. Understand that every fold serves a purpose, and each decision should be deliberate.

Identify Your Business Canvas

Just as origami artists start with a blank sheet of paper, begin with a clear canvas of your organization's strengths, weaknesses, opportunities, and threats. Define your vision, mission, and values—the cornerstone of visionary leadership.

Fold with Purpose

In origami, each fold has a purpose—to create a masterpiece. Apply the same principle to your strategies. Ensure that every action, every pivot, contributes to your overarching vision.

Adapt and Evolve

Origami creations often evolve during the folding process. Similarly, your strategies should be agile and adaptive. Monitor the business landscape, gather feedback, and be prepared to pivot.

Collaborate and Share

Origami enthusiasts often share their creations, learning from each other. Encourage collaboration and knowledge sharing within your organization. Foster an environment where innovative ideas can flourish.

The Role of Leadership

Leadership, in the context of Business Origami, is not about hierarchy; it's about influence. You are the origamist, guiding the folds, shaping the future. Here's how you can lead effectively:

Lead with Vision

Set a compelling vision that inspires your team. Just as origamists envision the final form, your vision should be crystal clear, driving your organization forward.

Empower Others

Much like origami artists who might work in teams, empower your employees to contribute their ideas and creativity. In the world of Business Origami, everyone can be a folding artist.

Embrace Change

Change is inevitable, and in the world of origami, it's where transformation happens. Similarly, embrace change in your organization. Be open to new ideas, pivot when necessary, and encourage innovation.

The Origami of Innovation

Innovation is the beating heart of Business Origami. Just as origamists innovate with folds, your organization should continuously seek innovative solutions. Here's how:

Foster a Culture of Innovation
Much like origami requires a creative environment, cultivate a culture where innovation thrives. Encourage brainstorming, experimentation, and the freedom to fail.

Adapt to Customer Needs
Origami artists often customize designs to suit their preferences. Similarly, stay attuned to your customers' needs. Use data and feedback to tailor your products or services.

Leverage Technology
Origami artists have adopted technology for new possibilities. In the digital age, embrace technology for efficiency, data-driven decision-making, and automation.

Sustainability: The Final Fold

Just as the final fold in origami completes the creation, sustainability should be the ultimate fold in your business strategy. Consider the environmental,social, and ethical impact of your actions. Embrace the principles of sustainability, as they align perfectly with the elegance of origami.

As we have reached here of our journey through the world of Business Origami, remember that you are not just a reader; you are now an origamist of strategy. The principles you've learned—the visionary leadership, strategic agility, innovative thinking, adaptive culture, customer-centricity, resilient operations—are the folds that shape your organization's destiny.

Embrace the endless possibilities of Business Origami. Just as an origamist can transform a simple sheet of paper into intricate art, you can transform your organization into a masterpiece of success. The canvas is yours, the folds are your choices, and the possibilities are endless. So, go forth and create your origami-inspired future—one fold at a time.

Your Path to Unfolding Success

In the preceding chapters, we've embarked on a transformative expedition through the intricate folds of Business Origami. We've delved into visionary leadership, strategic agility, innovative thinking, adaptive culture, customer-centricity, resilient operations, and the endless possibilities that await those who dare to wield this artistry in the realm of business.

Now, it's time to bring the focus back to you—the reader, the leader, the origamist of your organization's future. This chapter is not just the culmination of our exploration; it's a map to guide your journey toward unfolding success.

Visionary Leadership Redux

As we've learned, visionary leadership is the compass that sets your direction. It's about having a clear vision, articulating it, and inspiring others to follow. In the world of Business Origami, leadership is not confined to the boardroom; it's distributed throughout your organization. Every team member can be a visionary leader, contributing to the collective vision.

Your Path: Begin by revisiting or crafting your organization's vision. Make it a beacon that guides every fold of your strategy. Foster a culture where every team member understands and aligns with this vision, regardless of their role.

Strategic Agility: The Ever-Ready Navigator

The art of Business Origami teaches us that strategic agility is not just about reacting to change; it's about proactively shaping it. Your organization should be nimble, capable of pivoting and adapting in response to shifting market dynamics. Here, agility is not a buzzword; it's a mindset deeply ingrained in your culture.

Your Path: Start by assessing your organization's current agility. Are your processes flexible? Do you have mechanisms for rapid decision-making? Encourage a culture of experimentation and risk-taking. Your strategy should be a living document, ready to fold in response to emerging opportunities and challenges.

Innovative Thinking: The Creative Fuel

Innovation is the lifeblood of Business Origami. It's about finding new ways to create value for your customers and stakeholders. Innovative thinking should permeate every corner of your organization, from product development to customer service. Your business should be a hotbed of creativity, where every team member is encouraged to explore fresh ideas.

Your Path: Encourage innovation at every level. Create platforms for idea-sharing and experimentation. Promote cross-functional collaboration, where diverse perspectives come together to spark innovation. Remember, innovation isn't just about technology; it's about delivering more value to your audience.

Adaptive Culture: The Resilient Core

Adaptive culture is the foundation on which Business Origami rests. It's a culture that thrives on change, sees it as an opportunity, and responds with resilience. It's a culture where your team members feel empowered to adapt and innovate in their roles. In this culture, change is not feared; it's embraced.

Your Path: Building an adaptive culture is an ongoing journey. It starts with leadership. Lead by example, showcasing your own adaptability. Communicate the importance of change, and recognize and reward those who embrace it. Foster a culture of learning, where every challenge is an opportunity to grow stronger.

Customer-Centricity: The North Star

Your customers are at the heart of your origami-inspired journey. Just as an origamist customizes their creation to delight the recipient, your organization should obsess over understanding and satisfying your customers' needs. Customer-centricity is not a department; it's a philosophy that should permeate every action.

Your Path: Begin by truly understanding your customers. Invest in research, gather feedback, and create customer personas. Map the customer journey and identify pain points and opportunities for improvement. Train your teams to empathize with customers and always seek ways to exceed their expectations.

Resilient Operations: The Unwavering Foundation

Operational resilience is the backbone of Business Origami. Just as a well-constructed origami piece can withstand time, your organization should be built to weather storms. Resilience is not just about recovering from setbacks; it's about thriving in the face of adversity.

Your Path: Start by identifying vulnerabilities in your operations. Where are you most exposed to risks? Develop comprehensive contingency plans. Invest in technology that enhances your operational resilience. Most importantly, ensure your teams are equipped with the skills and mindset to navigate challenges.

The Journey to Success

As we reach the end of this book, remember that your journey is just beginning. Business Origami is not a destination; it's a way of navigating the ever-evolving landscape of business. The principles you've encountered here are your compass, your toolkit, and your guide. As you continue your journey to unfolding success, remember that your actions inspire those around you. You're not just shaping the destiny of your organization; you're setting an example for others to follow. Your origami-inspired strategies have the power to influence and transform not only businesses but entire industries.

You are now the origamist of your organization's future. Every fold, every crease, and every strategic decision you make shapes the destiny of your business. You have the power to create something extraordinary—a masterpiece of success, innovation, and resilience.

In your journey to unfolding success, there will be challenges and uncertainties, but that's where your newfound skills shine brightest. Just as an origamist embraces the complexities of their art, embrace the complexities of your business. Embrace change, adapt, innovate, and, above all, lead with vision.

The possibilities are endless, and the canvas is yours. As you embark on this journey, remember the words of the ancient Japanese art of origami: "Fold, but never break." The world of business may change, but your commitment to excellence, creativity, and growth will remain unwavering.

So, go forth, origamist of business, and create your masterpiece. The world is waiting to see what wonders you will unfold.

Thank you for joining us on this incredible journey through the world of Business Origami. May your every fold be a masterpiece, and may your future be as intricate, beautiful, and meaningful as the folds you create.

......The End......

A Heartfelt Thank You

I want to extend my deepest gratitude. Thank you for embarking on this journey with me through the world of Business Origami. Your commitment to learning and growth is an inspiration. As you continue on your path to success, always remember that the art of origami is not about the final creation; it's about the process, the patience, and the artistry. Likewise, your journey to success is not just about the destination; it's about the impact, the growth, and the transformation.

In the spirit of origami, may you continue to fold, adapt, and shape your future with creativity and purpose. And in doing so, may you find success that is as intricate and beautiful as the most exquisite origami creation.

Wishing you continued success,

Lloyd Jose Fernandez

Additional Resources and Tools

As you conclude your journey through the world of Business Origami, I want to ensure you have access to a range of valuable resources and tools to support your ongoing growth and application of the principles you've learned. Below, you'll find a curated list of books, courses, organizations, and tools that can enhance your understanding and implementation of Business Origami concepts (I am not associated or affiliated with any suggestions below).

Recommended Reading

1. "The Innovator's Dilemma" by Clayton Christensen: This seminal work explores why great companies can fail when they ignore disruptive innovations. Understanding the innovator's dilemma is key to strategic agility.

2. "Start with Why" by Simon Sinek: Sinek's book delves into the importance of visionary leadership and understanding the "why" behind your organization's existence.

3. "Design a Better Business" by Patrick Van Der Pijl, Justin Lokitz, and Lisa Kay Solomon: This book offers practical tools and methods for designing innovative strategies.

4. "The Lean Startup" by Eric Ries: Learn how to apply lean principles to innovate and create a culture of continuous improvement.

5. "Drive: The Surprising Truth About What Motivates Us" by Daniel H. Pink: Discover the psychology of motivation and how it can shape your adaptive culture.

Online Courses

1. Coursera: Explore courses on topics like leadership, innovation, and strategic agility from top universities and institutions.

2. LinkedIn Learning: Access a library of courses on leadership, creativity, and change management.

3. edX: Enroll in courses on adaptive decision-making, sustainable business practices, and more.

Professional Organizations

1. Business Agility Institute: Connect with a global community of agile practitioners and access resources on adaptive culture and strategic agility.

3. Association for Talent Development (ATD): Explore resources on leadership development and fostering a growth mindset.

Tools and Software

1. Trello: Use Trello boards to implement agile project management and foster a culture of adaptability within your teams.

2. Tableau: Leverage data visualization and analytics to support data-driven decision culture.

3. Miro: Collaborate visually with teams, perfect for design thinking and innovation workshops.

Networking and Communities

1. Meetup: Join local or virtual business and innovation-related meetups to network with like-minded professionals.

2. LinkedIn Groups: Engage with LinkedIn groups focused on leadership, sustainability, and innovation.

Your Journey Continues...

Remember, your journey doesn't end here—it's just the beginning. The principles of Business Origami are not static; they evolve, adapt, and respond to the changing world. As you continue to unfold your strategy and lead your organization, stay curious, stay innovative, and stay committed to the art of Business Origami.

References and Further Reading

In crafting "The Art of Business Origami: Folding Your Way to Success," I've drawn upon a wealth of knowledge, research, and insights from a diverse array of sources. These references and recommended reading materials have greatly contributed to the content and ideas presented throughout the book. Whether you're looking to explore these topics in greater depth or seek further inspiration, these resources will serve as valuable guides on your journey to business excellence.

Books

1. Christensen, C. M. (1997). **The Innovator's Dilemma: When New Technologies Cause Great Firms to Fail**. Harvard Business Review Press.

2. Sinek, S. (2009). **Start with Why: How Great Leaders Inspire Everyone to Take Action.** Portfolio.

3. Van Der Pijl, P., Lokitz, J., & Solomon, L. K. (2016). **Design a Better Business: New Tools, Skills, and Mindset for Strategy and Innovation.**

4. Ries, E. (2011). **The Lean Startup: How Today's Entrepreneurs Use Continuous Innovation to Create Radically Successful Businesses.**

5. Pink, D. H. (2009). **Drive: The Surprising Truth About What Motivates Us.** Riverhead Books.

Academic and Research Journals

1. Harvard Business Review

2. MIT Sloan Management Review

3. Journal of Organizational Change Management

4. Journal of Sustainable Development

5. McKinsey Quarterly

6. Deloitte Insights

References within the Book

Throughout the chapters of this book, you'll find references to various concepts, models, and real-world examples. These citations are meant to provide you with a deeper understanding of the topics discussed and to encourage further exploration.
I encourage you to delve into these resources and embark on your own journey of discovery and transformation. The pursuit of knowledge is a continuous process, and the insights gained from these materials can serve as powerful tools to help you fold your way to success in the ever-evolving business landscape.

Wishing you a future filled with innovation, adaptability, and sustainable growth.

Acknowledgments

Writing a book is never a solitary endeavor. It is the culmination of the collective wisdom, support, and inspiration of many individuals and resources. I would like to take this moment to express my heartfelt gratitude to those who have been instrumental in bringing "The Art of Business Origami: Folding Your Way to Success" to life. Writing this book has been a journey of exploration, learning, and creativity, and I couldn't have done it without the support and inspiration of many.

First and foremost, I extend my deepest appreciation to my family for their unwavering support throughout this journey. Your patience, encouragement, and understanding have been my anchor in the sea of creativity and dedication.

I am grateful to the many thought leaders, business experts, and authors whose work has inspired and informed the ideas presented in this book. Your contributions to the fields of business, leadership, and innovation have laid the foundation upon which this book is built.

I'd like to acknowledge the academic institutions, organizations, and researchers whose work has informed and inspired the content of this book. Your commitment to advancing knowledge in various fields has been instrumental in shaping the ideas presented here.

I extend my gratitude to the readers who embark on this intellectual voyage with an open mind and a thirst for knowledge. Your curiosity and commitment to growth are the driving force behind the words on these pages.

To all of you who have played a part in this journey, whether through direct involvement or silent support, I am deeply grateful. This book stands as a testament to the power of collaboration and the potential for transformation in the world of business.

Thank you all for being part of this remarkable journey. May the concepts within this book serve as a source of inspiration and guidance as you navigate the intricate folds of business success.

With gratitude,

Lloyd Jose Fernandez